Human Resource Management in High Technology Firms

Human Resource Management in High Technology Firms

Editors

Archie Kleingartner
Carolyn S. Anderson
The University of California at Los Angeles

Institute of Industrial Relations, UCLA

WITHDRAWN

Lexington Books
D.C. Heath and Company/Lexington, Massachusetts/Toronto

Library of Congress Cataloging-in-Publication Data

Human resource management in high technology firms.

Bibliography: p.

Includes index.

Contents: Human resource management in high technology firms and the impact of professionalism/Carolyn S. Anderson and Archie Kleingartner—High technology labor markets/Richard S. Belous—The educational implications of the high technology revolution/Lewis C. Solmon and Midge A. La Porte—[etc.]

1. High technology industries—Personnel management—Congresses.
2. Industrial relations—Congresses. 3. High technology industries—Employees—Congresses.
I. Kleingartner, Archie. II. Anderson, Carolyn S. III. University of California, Los Angeles.
Institute of Industrial Relations.
HF5549.H875   1987   658.3   86-45556
ISBN 0-669-13686-7 (alk. paper)

Published simultaneously in Canada
Printed in the United States of America
International Standard Book Number: 0-669-13686-7
Library of Congress Catalog Card Number 86-45556

The paper used in this publication meets the minimum requirements of American National Standard for Information Sciences—Permanence of Paper for Printed Library Materials, ANSI Z39.48-1984. ∞ ™

87 88 89 90 91 8 7 6 5 4 3 2 1

Contents

Preface

Higch technology industry receives more coverage in newspapers and popular periodicals than any other sector of economic activity. There is no question that the United States has entered a period of sustained economic and social change as a result of the remarkable technical innovations associated with high technology firms. These changes are substantially transforming the nation's industrial structure and creating thousands of jobs whose titles were not even invented a few years ago. They are greatly accelerating the rate of product innovation and diffusion, will alter how people live and work, and may well reposition the role of the United States in the global economy.

A popular mystique has developed around the creative geniuses who make the initial technical breakthroughs and the entrepreneurs who make and market the products. Application of these innovations on a wide scale—their design, production, and distribution—depends, in the final analysis, on large numbers of people, within organzations, performing interrelated functions. It is people who sustain new firms, create new industries and new markets, and shape the overall impact of these innovations. To get the results of technical innovation into commercial use requires organization and a cast of support personnel—engineers, technicians, assemblers, paper handlers, and managers. The ability of firms, industries, and, indeed, the United States as a whole to compete effectively in international markets over the long term hinges on a broad spectrum of human skills and on crucial organizational decisions affecting their deployment.

This book provides an introduction to key human resource issues faced by high technology firms and industries. The various chapters present informed analyses of these issues to help firms better assess their own approaches and practices. They also provide a basis for the establishment of human resource policies suited to the needs and circumstances of the particular firm.

This book does not offer cookbook approaches for designing particular human resource management systems and practices; many such books and services are already available. Rather, taking examples from the world of high technology industry, the book enables thoughtful executives and human resource managers in individual firms to place their firm's needs in perspective. It makes decision makers aware that there are many human resource options available to them and reminds them of the implications of such options over both the short and the long term.

Serious analysis of the human resource aspects of high technology development has been overshadowed by the glamor of other topics touted in the popular media—topics that often overlook the differences among firms in the high technology category. Indeed, the very question of what defines a high technology firm requires attention.

Although there are differences of opinion regarding what constitutes a high technology firm, it is convenient to think of it as a matter of degree: firms can be placed along a continuum, and a profile can be drawn of those that define the high technology universe. In general, the distinguishing feature of a high technology firm is that the conceptualization of new, technically based ideas and innovations and the commercialization of these new ideas are basic to its business strategy. A very broad definition of high tech thus includes a sizeable number of industries.

The industries in which criteria defining the degree of high tech are most likely to be found, as classified by the U.S. Bureau of Labor Statistics, are electronic components, communications equipment, guided missile and space vehicles, aircraft, drugs, and office computing and accounting machines. The emerging field of biotechnology and the explosive growth of information technologies will continue to enlarge these groups over time, as new fields develop and as established fields are revolutionized.

We may think of high technology industries as those that share these attributes:

The proportion of engineers, scientists, and technicians in their work force is higher than in other manufacturing industries.

They are science-based in the sense that their new products and production methods are based on applications of science.

Research and development are more important to their successful operation than they are to other manufacturing companies.

They depend on the academic community to educate their work force to a larger degree than traditional manufacturing companies do.

The markets for their products are both national and international.

The life of their products tends to be short, with products often becoming obsolete before mass production can be undertaken. (Western Technical Manpower Council of the Western Interstate Commission for Higher Education, 1983)

High technology industries have additional important characteristics that have a major impact on human resource management. They have job multiplier effects in sectors related to the distribution, sales, and services of their products.

They improve the productivity of traditional firms that use their products. They emphasize the demand for and utilization of highly trained professional workers.

A major feature of the chapters in this volume is the authors' recognition that human resource decisions are linked to the firm's relative stage in its own life cycle and that of its product or industry. In addition, outside the firm's control but affecting human resource management decisions are the environmental pressures of the labor market, the business cycle, government regulation, and international competition. All this has explicit relevance to the kinds of human resource policies the firm implements. Human resource practices that are recognized as innovative and successful in a firm at one time may be inappropriate at a later date.

Topics covered in this book include recruitment and retention of key personnel, compensation strategies, conflict resolution, solutions for skill obsolescence, and sharing of corporate goals and values within the work force. The book addresses such questions as how much difference there is between human resource management practices in high technology and more traditional firms and how these differences can be explained. It explores the ways in which these practices can adapt effectively as the firm and industry change.

The book also deals with the demands high technology growth places on educational institutions to prepare students to work effectively in the new environment and the reciprocal relationship between industry and academe. It looks at how high tech firms manage employee relations and what implications these personnel practices hold for labor unions. The features that characterize organization of work and production in high tech firms and their consequences for productivity and organizational effectiveness are also addressed. In this regard, the book raises the difficult issue of determining how varying human resource policies support productivity in the firm's various jobs.

This book is directed toward three principal audiences. The first audience is managers and human resource practitioners in high technology firms who wish to gain insight and perspective about their own human resource problems and practices. Effective human resource management does not exist in a vacuum, and high tech firms—perhaps more than most organizations—are vulnerable to the myopia that can come with the currently abounding hype and headlines. A second audience is composed of managers and human resource specialists in more traditional firms and industries that are affected by what is going on in high tech. Some are affected directly—for example, firms that are now absorbing high technology organizations or firms that are moving from more traditional areas into high technology activities and want to be aware of modifications of human resource strategies in order to adapt. Others are indirectly affected but may be aware of and interested in innovation and change that can influence the productivity of their own work forces. Third, the book should find a place in classrooms as supplementary reading in courses concerned with human resource management and industrial relations where the emphasis is on understanding the forces that will shape human resource practices in the coming years.

Although the book's focus is on human resource issues associated with high technology development, the implications of these issues are not limited to high technology organizations in the narrow sense. Those who are interested in the general issue of how human resource utilization shapes the creation, growth, and demise of organizations will find relevant material here.

One of the strengths of this book is the expertise of its contributors; all are persons whose work is well known. All of them are enthusiastic contributors, and they recognize that the issues and innovations of human resource management practices of high technology industries have implications for other industrial sectors. Besides contributing their individual chapters, the authors came together for a one-day seminar at the University of California, Los Angeles, to present their findings and to discuss reactions from prominent educators and business and human resource specialists.

The first chapter, prepared by the editors, explores the genesis of popular perceptions of human resources in high technology firms. There is a widespread belief, for example, that high technology generates a distinctive work place culture—that high tech firms are characterized by a new breed of worker, new approaches to the organization and supervision of work, and new incentive systems. The sources and rationale of this culture are explored because, despite a basic validity and widespread appeal, these images are a vast oversimplification of the real situation in most high technology organizations.

Chapters 2 through 10 are grouped into parts that deal with the high technology work force, human resource concepts and practices, and the implications for industrial relations. In a general way, these three parts cover the external determinants of human resources in high tech firms, the utilization of these resources within organizations, and the industrial relations consequences of the interactions that occur within these work organizations.

The final chapter reviews the major themes covered in the book and incorporates comments from a select group of government, academic, labor, and management experts. This chapter also provides perspectives regarding the significance and relevance of changing human resource management practices in high tech as firms and the industry mature.

Acknowledgments

The research reported in this book was funded by the Institute of Industrial Relations at UCLA. In particular, we wish to acknowledge the support and encouragement of Daniel J. B. Mitchell, director of the Institute, throughout the project. Jane Abelson Wildhorn, administrator of the Publications Center at the IIR, provided invaluable editorial commentary and assistance. Ms. Wildhorn was ably assisted by her staff—in particular, Sharon Geltner, Margaret Zamorano, and Jeannine Schummer. Judith Richlin-Klonsky, a doctoral student in sociology at UCLA and an IIR research assistant, made a particularly important contribution during the research phases of the project. Lynne C. Zucker, associate professor of sociology at UCLA and a research associate of the IIR, provided helpful background material and insightful comments on several of the chapters. Victor Tabbush, associate dean of the Graduate School of Management and director of its Executive Education Program, helped us organize a survey of high tech executives. Administrative assistance was provided by Patricia Kracow.

We also owe a debt to Professor Umberto Sulpasso for organizing a conference in Bari, Italy, where Karl Pister first presented the material contained in chapter 4. And we wish to thank Bruce Katz, our editor at Lexington Books, for his encouragement and support.

Last but not least, we wish to acknowledge the support we received from our respective families: Dorothy, Calvin, Jane, Elizabeth, Garrett, Posie, Molly, and Thomas. Some contributed more than others, but we love them all.

Part I
Introduction

1

Human Resource Management in High Technology Firms and the Impact of Professionalism

Carolyn S. Anderson
Archie Kleingartner

The remainder of this century is frequently viewed as the period during which we will see widespread diffusion of the results of high technology innovation into the lives of ordinary citizens. It is not necessary to believe that there will be a personal computer in every household and on every elementary school student's desk to appreciate the pervasive impact of these developments. The impact of the new technologies can be readily appreciated by most adults when we recall that ten years ago there were few, if any, automatic bank teller machines and no scanning markers (instead of price tags) on supermarket items and that a walk through a factory still conveyed the feeling that people, not machines, were doing the finish work.

Development of high technology industry is serious business. It is also, in many quarters, a political football—a fad replete with opportunists eager to seize on the confusion and uncertainty that surround many aspects of its development. Bookstores and newsstands are filled with titles and headlines intended to shape our perceptions of high technology. Overshadowing the serious analyses and commentary is a plethora of material promoting the quick fix, with easy answers to almost every problem.

Questions about the human resources required to manage, produce, distribute, and carry out the myriad other activities associated with the results of high technology innovation are not exempt from these tendencies. It takes time for information about technological innovation to diffuse and become comprehensible. It also takes time for knowledge to accumulate about the human resource ingredients that go into the creation, manufacture, and distribution of these innovations.

When mass production originated in the automobile industry and the assembly line first came into widespread use, it was thought to be not only the most efficient method of production from an engineering standpoint but also highly conducive to the development of good employee relations. History has demonstrated that these beliefs fell short of the mark on both counts, although they have had a pervasive impact. The classic assembly line method of production has been in the process of modification for many years and is rapidly giving way to new modes of production. The dehumanization and dissatisfaction associated with work on the assembly line has been documented in many surveys and studies. But in its early period, the assembly line was touted as signifying a major breakthrough in the organization of work and workers.

Today Silicon Valley, Route 128, and their progeny conjure up similar state-of-the-art images of a technological and economic frontier that will set the tone for business activity and human resource management well into the twenty-first century. Hoping to venture aggressively into the high tech frontier, virtually all of our fifty states have established special state agencies charged with creating their own versions of Silicon Valley. Likewise, state departments of economic and business development broadcast materials to attract high tech to their area; multipage advertisements in *Fortune* and *Business Week* appeal to tenants for "Silicon Mountain," "Silicon Prairie," "Silicon Forest," and the like. And for all those in government and business who are concerned with the ability of American high technology to compete in international markets, there is a best-seller almost every month that presents a theory for solving managerial and competitive problems relative to work in the industry.

Many people argue that technology is creating a new information society, in which knowledge and information form a basic resource which will radically transform the nature of work; they claim that we are in the midst of the transition from an industrial society to an information society. However, the futurists, who are concerned with the end results, often convey the impression that they are dealing with the contemporary scene; in the process, they create widespread anxiety in various quarters about being left behind.

The allure of the developments in high technology includes the feeling that vast entrepreneurial opportunities await those with the foresight and talent to take advantage of them immediately. The rate of product innovation and diffusion in these industries is accelerating, and there is an equally accelerated rhythm of change and adaptation in the organizational and human resource management practices in high tech firms. It is widely believed that many such companies are using innovative human resource management strategies that maximize the productivity of their work forces. The context in which such strategies are initiated, adopted, and—as often as not—eventually discarded is not so clear.

What is occurring in high technology firms is likely to have large ripple effects in other industries. This happens, for instance, when a traditional firm buys out a successful entrepreneurial high technology company. It also happens when high tech firms move out of the cloister of Silicon Valley and its progeny to

Wichita, Omaha, Cleveland, and other locales that are eager to get a piece of the high technology action. In these circumstances, the high technology firms become role models for other organizations in the area to a greater degree than their size alone would suggest.

Because of high tech's innovative character and the widespread effects of its distinctive strategies, the human resource management issues examined in this book have significant implications for policymakers, educators, and the institutions of the work place. Any one of its topical chapters can be read separately on a descriptive level for information about the high tech labor force, education and skill training at school and on the job, recruitment, managing change, compensation strategies, grievance systems, implications for unions, and the like. However, the book is more than description. There is embedded here a depth of perspective that clarifies a number of stereotypical perceptions about high tech.

This perspective addresses the origin and structure of innovative human resource management practices in the industry as well as the ways in which such practices are most productively linked to a firm's business strategies and the external social, economic, and political realities that affect them.

The Promise of High Tech

Glamor and hype about high tech's potential in regard to jobs and the economy still permeate its popular image. The rosy glow that surrounds high tech is not limited to the economic benefits the industry is presumed to ensure for workers and communities. No small part of the lure is the expectation that high tech means new and better kinds of work for all involved. And high tech workers themselves are perceived to be a new breed—different in motivation, with high levels of scientific and technical expertise, and more educated and professional than those who work in other more traditional industries.

The ways in which the work is done are believed to be different, too. New corporate philosophies are said to dictate organizations without the usual bureaucracy and hierarchical supervision and with guarantees for workers that they will have the challenge and the opportunity to accomplish meaningful and important work. Work in high tech is thus touted, above all, as intrinsically rewarding and fulfilling.

Compensation strategies are seen as directly linking workers to their companies' financial performance, providing clear incentives that not only motivate them to higher productivity but also build commitment and loyalty to the firm. Such incentives to high and increased productivity by the high tech work force are overwhelmingly emphasized by the press. Enlightened management practices are presumed to provide workers the same protections and guarantees that unions have traditionally sought for their members. According to this rationale, unions don't exist in high tech because they are not needed.

Even production work in high tech is portrayed as different from production work in other industries. It is clean work, according to early reports about the industry. Hazardous working conditions and air pollution are relics of the smokestack era. Magazine photos portray high tech production work as not only more scientific but also more antiseptic than other production work.

The Paradox of High Tech

It is not surprising that a number of recent critics have refuted these glowing images of high tech on several grounds. Their criticisms challenge those who want to reap the promises of high technology without fully understanding its complexities and limitations.

We feel that any discussion of human resource practices in high tech firms has to acknowledge that what the media highlight, and what is intriguing about high tech, is only part of the story. Although high tech will continue to grow, expansion of employment will be constrained by a number of factors, as we will discuss. Not all high tech workers are motivated, highly trained scientists, engineers, and technicians. Not all work is challenging and autonomous. In many work places, especially those involved in production of commercial products, work can be repetitious and routine. Although wages and other compensation for professional and managerial workers are often high, wages for some portions of the high tech work force are among the lowest in manufacturing industries, reflecting very real differences among the skill levels of workers in the industry. And there are few reliable estimates of the prevalence or effects of innovative compensation and other human resource management practices in the industry.

High tech products are subject to cyclical demand and thus the industry is unstable with respect to employment levels (Burgan, 1985). There are more part-time employees in high tech—and, hence, fewer desirable full-time jobs—than in other manufacturing industries (Anagnoson and Revlin, 1985). Unemployment rates are often high. Business failures are more frequent (Freeman, Carroll, and Hannan, 1983). There is little evidence that management in hard-hit firms sees the utility of such enlightened employee relations practices as guaranteed jobs.

Finally, working in high tech industry can be hazardous, according to public health officials. The dangers from poisonous chemicals used in some manufacturing practices have been acknowledged and are the target of remedial activity by public agencies and within the industry.

Before turning to our own analysis of the genesis of much human resource management innovation in the industry and to the research findings of our other authors, we wish to address a number of beliefs about high tech, to explore how many of the practices that they highlight seem to have evolved, and, in that regard, to outline some unaddressed issues and questions that are raised by the findings in this book.

The Employment Factor

There has been a common perception that employment in high tech industries can make up for losses and lack of growth in the traditional industrial manufacturing sectors. State governors have been the most notable enthusiasts of high tech as the harbinger of the nation's reindustrialization, and their optimism continues to underwrite those who see high tech as an endless generator of new jobs.

Despite the fact that high tech is expanding and that new jobs are being created, their absolute numbers are modest, according to Richard Belous (chapter 2). In addition, what jobs there are do not resemble lost jobs in terms of either skills or pay. The Office of Technological Assessment and numerous JTPA training programs across the country have tried in vain to link the skills of structurally unemployed workers in the traditional industries to projected demands for skills in high tech. The skills just don't fit.

Belous points out that even the most inclusive definition of high tech sets current employment at 13.1 percent of the total civilian work force, with the most optimistic projections to 1995 for 14.1 percent. Although projections indicate that the rate of growth of high tech has been high since 1972 and should continue to outpace that of other industries, it will account for only 17 percent of new jobs at best through 1995. (More optimistically, such projections do not include the indirect effects of high technology industry on jobs, a subject that Belous does consider. And since much high tech development has been geographically concentrated in some areas, such as Santa Clara County, California, the impact of high tech on employment is pronounced.)

Also, as noted earlier, much of the industry is characterized by economic instability and unstable employment levels. Currently, high tech is said to be going through a shake-out period because of excess capacity and temporary market saturation as well as international competition. Firms with long-standing policies against layoffs have laid off workers. Some firms have assigned unpaid vacations for the first time. Other firms have merged, with immediate effects on organization and work force size, or have disappeared altogether. Enthusiasm for high tech as ensuring more jobs—as economic salvation for regions beset by unemployment—should be tempered, at least for the short term.

The Offshore Phenomenon

High tech jobs are also subject to export. The growth of the offshore phenomenon in high tech is parallel to the runaway shop syndrome in other industries, in which various stages of production are exported to areas that have lower production costs, primarily because of wage differentials. The significance of the offshore movement of the industry will not be limited to the displacement of unskilled production jobs from the United States. Allen Scott (1985a, 1985b) notes that the displacement of low-skilled jobs also leads to increased displacement of related higher-skilled jobs.[1] Whether this displacement will continue or reach equilibrium will depend not only on the worldwide growth of high tech but also on the relative U.S. share of the industry.

Our understanding of the demand for production jobs in the U.S. is clouded by the movement of such jobs offshore. Within the United States, production jobs still comprise the majority of high tech jobs. So long as most production took place within the country, it was easy to estimate, at aggregate levels, the proportion of any particular industrial work force that was composed of production workers, as well as their skill levels, and thus to predict demand for future workers. In the high tech defense industries, this is still largely true because of federal regulations that insist upon domestic production (although these industries are notoriously volatile for other reasons). It is no longer so reliable to estimate future demand for production workers in those U.S. high tech firms for which most parts of final products are made elsewhere in the world, although they may be designed, packaged, or marketed here at home.

There is no reason to think that the trend toward overseas production will stop. Almost every month, one electronics firm or another announces the opening of a new production unit in Korea, Taiwan, or Brazil. Moreover, according to Scott (1985b) there has been no indication that increasing mechanization of production will result in the repatriation of mechanized jobs to this country.

It is difficult to estimate total employment levels in high technology industries that cross national boundaries, because of the nature of the skill distribution of the work force. We know that high percentages of professional and technical workers work within the country and that unknown proportions of the unskilled production work force work abroad. If the present situation continues and high tech continues to grow, the demand for professional workers should continue to be high, independent of the demand for production workers.

However, even with substantial continuing export of production jobs, since product development includes day-to-day communication between the scientists and engineers who design the product and the production people who get it out, it seems unlikely that U.S. production work will be cut back completely. Whether there will be an increasing demand for unskilled high tech labor here relative to that for the skilled work force depends on the extent to which unskilled production jobs continue to be exported. Unskilled high tech workers and human resource management practices that apply to them get relatively little press coverage, however. The projected demand for low-skilled jobs—typically allotted to female, immigrant, and minority workers—is not nearly so media-marketable as the projected demand for workers with another image.

We turn now to a discussion of this other, more compelling image, which represents an important segment of the high tech work force and one that has had a dominant effect on innovative human resource management practices in the industry.

A New Breed of Worker—A New Kind of Work

The perception that high tech workers are a new breed seems, at first glance, to have to do with reflections of worker discontent such as those in *Work and the Quality of Life* (O'Toole, 1974)—the idea that American workers had new demands for participation, autonomy, and control in the work place. These demands, based on workers' psychological needs for fulfillment and self-actualization in their work, were linked to productivity improvement. Experiments with worker participation programs in the United States, Sweden, and other industrialized nations rekindled consciousness of the importance of ensuring worker involvement and integration in the major goals and purpose of industry. Not since the days of the Hawthorne experiments in the 1930s had so much management attention focused on issues emphasizing the necessity for integrating the purpose, the technology, and the social organization of work.[2]

Despite mixed evaluations, a broadly based movement to improve the quality of working life by extending opportunities for worker participation has by no means exhausted itself in traditional industries. At the same time, rising educational levels and shifts from blue-collar to white-collar work have affected the expectations that employees have in regard to jobs and careers. It is in this milieu that innovative human resource management practices attributed to high tech have evolved.

The new breed of worker is found not only in high tech, of course. The entire work force is changing. The blue-collar worker, whose work in the foundry, the steel mill, and the factory was always physically demanding, is being replaced. Even the white-collar worker, whose work was often unchallenging and repetitious, is being upstaged. Kelley (1985) tells us that there is a new "gold-collar" worker, who by 1990 will make up some 60 percent of the entire U.S. work force. Whatever the industry, gold-collar workers are, first and foremost, knowledge workers. Brains, not muscles, are what count. In an information society, as ours is frequently labeled, it is knowledge that supplies the economy with its most critical resources of production. And of all industrial sectors, it is high tech that the popular press identifies with the new breed of educated, innovative, autonomous workers. Their work is self-paced, their goals internally defined, their motivation unquestionably high and task-oriented.

The most salient characteristic distinguishing these highly publicized workers and their work in high tech is, in one word, professionalism. We believe that if there is, indeed, anything new and different about human resource management in high tech, it stems directly or indirectly from this characteristic of the best-known segment of the high tech work force and the ways in which its work is done. "Professionalism" thus refers to the skill level of engineering, science, and technical workers (their education and training), to the type of work they do, and to the ways in which they do it.

These workers share the scientific mystique that has cycled through the public consciousness since Sputnik catapulted math and science expertise into the nation's top human capital resources. As we note in the Preface and as Belous (chapter 2) shows, the proportionately higher employment of highly educated scientists, engineers, and technical workers is a criterion of almost all definitions of high technology industry. These concentrations of professional workers have been responsible, we believe, for the development of many of the distinctive human resource management practices and activities mentioned in this book: flexible compensation practices (with emphasis on firm performance and individual incentives in addition to competitive base salaries), dual career lines, job security guarantees, and corporate training programs. The need to maximize professional productivity has also led to the adoption or development of methods of work organization that limit hierarchical levels of supervision, increase professional commitment to the task and to the firm, and link performance evaluation and dual career lines. Furthermore, the dominance of concerns relative to managing professionals has helped influence corporate cultures, which extend such practices to employees at varying levels throughout the firm. Foulkes, Milkovich, and Miljus and Smith (chapters 5, 6, and 7) link the strategic importance of the firm's professionally educated work force with significant new human resource policies and practices. These new and not-so-new practices may be integrated into a "new" system of human resource management, which is discussed by Kochan and Chalykoff (chapter 10).

Many of the practices noted in high tech firms (although by no means limited to high tech) derive in no small part from the following management needs:

1. To recruit the professional segment of the high tech work force in the face of what have often been very tight labor markets.

2. To maintain professional employees' commitment to the firm.

3. To provide them with incentives and job security.

4. Above all, to foster their productivity in the innovation and development of high tech products.

Worker Productivity in High Tech

Ensuring professional productivity is the driving force behind the development of the new human resource management systems in high tech, for productivity is the demand that sets the need to recruit, motivate, facilitate, and reward the firm's strategic workers most effectively.

Traditionally, productivity in manufacturing industries has been evaluated in terms of tangible output. It is much more difficult to track and evaluate and thus to "manage" work that involves the production and development of ideas, especially at preproduct stages. Success is not automatic; sometimes it is not even probable. According to Dornbusch and Scott (1977) the nature of the work of innovation and development precludes the use of the traditional scientific management organization of work, in which the predictability of tasks is high, outcomes are measureable, and procedures are well established.

The work autonomy and independence that professional workers seem to enjoy and that are touted as integral aspects of the way things are done in high tech can thus be linked to the nature of the work done. Scientists and engineers in high tech need sufficient autonomy to respond to the uncertain or changing situations that confront them in the course of their work, to follow trains of thought and methodologies that may be productive, and to exclude others that are not. Such work is accomplished most efficiently under a management system that allows for more autonomy, fewer rules, less minute specialization, and less centralized decision making—one that provides flexibility in the means by which tasks are set up and accomplished. In general, evaluative standards for high tech science and engineering professionals must take into account the difficulties inherent in tasks for which workers do not have full control over the results of their efforts and (as in many high tech projects) when more than one worker has contributed.

The Sources of Professional Expectations

Professional workers are socialized into the ways in which they will work by their education. The education of college-bound youth and their subsequent college and university education in the United States is oriented toward individual achievement, independently accomplished. The expectation is that the individual will learn to tolerate ambiguity, to take the initiative, to solve problems—that he or she will internalize standards of expectation and performance and will do so without the sort of supervision that has characterized traditional production work in industry.[3] School and college work is evaluated according to standards of knowledge that are established and maintained by those within the discipline or profession.

Thus, professionals are socialized by their educational backgrounds to expect to work independently, without monitored and routine performance standards, and to work under conditions of challenge and autonomy. Performance is a response to internalized motives, which dictate goals as well as standards. And professionals expect their technical work to be evaluated by those who, besides sharing organizational titles and authority, share their technical backgrounds and expertise.[4]

Professionals and Industry: Conflicts

Although the concentration of scientists and engineers in high tech industry is the primary defining feature of high tech, the absorption of scientists and engineers into industry per se is not new. There has been a strong demand for such professionally trained workers in the aerospace and electronics industries as well as in certain other industrial R&D divisions and firms since World War II.[5]

By the early 1960s, the flow of these professionals into industry had attracted the attention of industrial relations and organizational theorists, who noted several key areas of potential conflict between professional needs and expectations and industry's traditionally bureaucratic organization of work.[6] Many of the developments in human resource management mentioned in the following chapters address these conflicts directly.

In those earlier days, problems were identified in the areas of goals, incentives, and sources and nature of authority. Because professionals' goals might be different from or conflict with those of the firm, ways were sought to accommodate such goals so that they complemented those of the company. In addition to economic incentives that brought commitment to the firm, professionals were thought to require incentives that addressed their professional expectations for accomplishment and status in their fields. Advancement in bureaucratically organized firms was typically via promotion into management. For scientific professionals in industry, this often meant the end of a career in professional specialization; for many, this was a difficult choice.

Traditionally, supervision in industry was by managers trained in management techniques either by experience or schooling. Their authority derived less from how well they knew the craft or the trade than from their authority level in the organization. With the increasing recruitment of professionals into industry—and since professionals were schooled to respect authority based on shared expertise, rather than job title—the supervisor trained in management techniques lacked credibility.

A manager or administrator of research, besides having a supervisory job title, had to be professionally trained. A manager who lacked the ability to evaluate the technical adequacy of work performed or any of the scientific or technical criteria involved, and who managed by virtue of a title alone, would not be effective, according to this rationale.

Unionism among Professionals

A good deal of interest has focused in the past on the possibilities for union organization of professionals who work in industry. Those who have seen unionization as a solution to the resolution of conflicts inherent in industry's professional–management relationship have tended to follow one of two approaches. The first emphasizes that professionals' goals, incentives, and expectations for control and evaluation differ from those of blue-collar workers in industry; therefore, any potential unionism would have to accommodate these concerns (see, for example, Strauss, 1963; Kleingartner, 1973).

The second approach notes that the work of many engineers and scientists in industry is more like the routine, technically skilled work of a craftsman or artisan than like that of an independent, autonomous professional (see, for example, Goldstein, 1955; Kornhauser, 1962; Strauss, 1963; Dvorak, 1963; Gordon and Ross, 1962). Many industrial research tasks do not demand autonomous decision making or flexibility, although technical skill is needed. The work is amenable to the methods of evaluation of traditional industry, which were based on following set procedures and measuring tangible output. If these similarities were pointed out, such "professional" workers would have interests much in common with blue-collar workers. Collective bargaining over terms and conditions of such work therefore seemed to provide a reasonable approach to employee relations issues, including job security and due process as well as wages. What was necessary, according to these analyses, was that the workers involved redefine their professionalism and acknowledge their commonalities with the needs and solutions of other workers.

Many early efforts to unionize such groups of engineers failed. Even when they have been successful, organizing drives of scientific and technical professionals in industry have seldom recruited large percentages of those represented to membership. Nor does unionism for professionals attract much interest in high tech industry today.[7] Everett Kassalow (chapter 9) found no indications of significant organizing of professionals by the AFL-CIO or independent unions. Managers surveyed indicate that unionization, whether of professionals or production workers, is not of much interest or concern.[8]

Specialization within the Professional Work Force

Although the issue surfaced years ago, specialization remains a salient phenomenon today, with implications for differential practices in regard to professional recruitment, training, and compensation strategies. In high tech, just as in other industries with substantial numbers of professionally trained scientific and technical employees, there is increasing specialization of the research function within the firm as it grows. Separate departments or divisions isolate basic research, applied research, and product development and servicing activities. The continuum of such specialization runs from the autonomously defined and accomplished work of basic research to the concrete technical tasks of large numbers of entry-level engineering and science graduates. The implications of this trend tend to go unnoticed. So does the concomitant division of labor within the research work force, yet such differentiation in work raises issues regarding differentiation in human resource management policies.

Most young recruits today, just as Gordon and Ross noted in 1962, are hired to perform the "mountains of routine skilled work which has to be done between the conception and fruition of scientific breakthroughs and pioneering engineering achievements." There is limited professional mobility for the majority into the rarefied air of the basic research units, although there is some promotion from entry-level engineering and technical jobs into applied research units and administrative jobs. For many, however, improvements in job conditions and income mobility come through switching employers. The long-term effectiveness of this strategy has yet to be shown. Apparently, despite interemployer mobility, life cycle earnings curves of most professionals in industry typically peak early and remain flat for the remainder of their careers, unlike those of many other professional workers outside industry.

Professionals and Industry: Accommodations

Organizational theorists in the 1960s and 1970s, though aware that industrial professional–bureaucratic conflict resolution via unionization was a real issue, often concentrated on the dynamics of what was occurring—the adaptation of the organization to the professional, and vice versa, so that development in aerospace, computers, or semiconductors could proceed apace.

And, although analyses such as those of Kornhauser (1962) and others pinpointed troublesome areas, there were indications that in the real world, productive efforts at accommodation were synthesizing organizational and professional goals, incentives, and forms of control and evaluation. Wilensky (1964) had noted the onset of the merging of organizational and professional interests: "The culture of bureaucracy invades the professions; the culture of professionalism invades organizations" (p. 321). Many of the developments currently described as new approaches to human resource management were influenced by these dynamics. We turn now to a brief discussion of several of the more significant organizational and human resource innovations presently in use among firms employing professional scientists, engineers, and technicians.

Careers and Compensation Systems

The identified conflicts between professionals and industry regarding incentives targeted the lack of opportunity for career growth for professionals within a firm, except by entering management positions (see, for example, Kornhauser, 1962, and the discussion by Kanter, 1985). The career and income mobility available to professionals who wanted to continue to practice their professions came through a change of employers. As noted earlier, although such mobility does exist, the effects on income are seldom significant beyond the early years of the professional's career.

Furthermore, this job changing proves dysfunctional to the firm, which has a considerable investment in the professional's career in recruitment, training, and replacement costs. It was therefore logical for firms to begin to develop, within their internal labor markets, opportunities for career growth by establishing dual career lines, which provide for income and status advancement for professionals parallel to that of management. Today, many high tech firms follow this practice (Kanter, 1985).

Commitment to the firm is also strengthened by the use of additional long-term and short-term incentives, such as those discussed by Milkovich (chapter 6). A typical firm's use of such incentives is described by Foulkes (chapter 5).

The Matrix Organization of R&D Work

The need for increased formalization and bureaucratization as a firm increases in size tends to go unremarked, since it is so natural a consequence of growth. Yet, as industry became more R&D oriented, rigidities of organizational structure within large companies often precluded the flexibility that research units seemed to demand. Increased utilization of the matrix organization of work was probably the most significant organizational response to professional–bureaucratic conflict (Smith, 1966). In this compromise approach, the professional has reporting relationships along two dimensions of a matrix: (1) to the administrator of the department into which he is hired and from which he is recruited to work with a team on research projects, and (2) to the administrator of the particular project. The professional is recruited from his home department into a particular research project. Such projects are of finite duration, a few months to several years. Once the project is over, the professional worker is reassigned to another project demanding his skills. The temporary nature of the project lends itself especially well to the suspension of hierarchical authority and replacement with more flexible problem-solving, task-oriented work organization and responsibilities (see, for example, Smith, 1966; Kanter, 1985).

The Project Administrator

Leadership/authority in such research projects is typically vested in a project administrator. Technical expertise ensures his credentials with those whom he supervises. The emergence of the role of the project administrator, or "scientist-administrator," resolves a number of professional–organization conflicts. It ensures mutual respect and deference to a body of knowledge as well as to ways of working on the particular research project. At the same time, most of the administrative work relative to the professional's career within the organization proceeds within the technical department into which he is recruited and continues along traditional bureaucratic lines, in which scientific expertise plays relatively little part.[9]

Such discussions imply the sudden arrival on the scene of a new (within the past few decades, at least) form of work organization and new qualifying criteria for managers—hence, the matrix and the research administrator. These developments, widespread as they now appear, seem to have occurred differently in different firms, depending on their age. In traditional firms, the gradual specialization and increasing significance of research activities may have resulted in debureaucratization within the firm's research divisions. At some R&D firms, there may have been fairly revolutionary implementation of organizational change (see, for example, Smith, 1966).

The history of many start-up high tech firms suggests a much more evolutionary scenario. In firms such as Hewlett-Packard, Fairchild, Intel, and Apple, the founders of the firm were most often highly educated, technically trained engineers and scientists who were also entrepreneurs—developers of the firm's original product. These corporate founders worked primarily on tasks relevant to the research project, in ways that reflected the demands of the work and their professional expectations about how it was to be done. Thus, in our survey of high tech executives, staff members trained in line management techniques (including human resource management) were hired considerably after the founding of the firm, and only when it had grown to the extent that such functions demanded full-time attention. The importance of the cutting-edge research project and the kinds of human resource management involved in handling professionals' concerns assumed lesser proportions, since other kinds of management decisions and activities became more important to the firm's profitability and continued growth.

As such high tech firms expand beyond the product development stage, the relative importance of the professional to the firm's activities is diluted by demands for managers and workers with other kinds of skills. The case of Steven Jobs's replacement at Apple by a management executive he had hired to implement necessary business strategies is an extreme example of a transition from a firm primarily responsive to R&D demands (with their link to professionalism) to a firm primarily responsive to the demands of the market.

Such transitions mean shifts in the skill level of the work force as well as in the kind of training and expertise most appropriate to management. With these shifts come changes in the expanded emphases of human resource management.

The Forgotten Majority

Although some of the popular literature suggests that all high tech workers, from the top down, are highly skilled, talented, and creative, analysis at the aggregate level (see Belous, chapter 2) shows that a sizable percentage of the high tech work force is composed of relatively low-skilled blue-collar workers. The 1980 U.S. Census of Population also shows that although the percentage of professionals and managers in various high tech industries may run very high, nonetheless the majority of the work force consists of production workers.

Patterns of the mix of workers involved in the manufacture of high tech thus display what Alejandro Portes (1985) suggests may be "the essence of modernity"—the Silicon Valley model of a work force, with a sharply defined two-tier skill distribution. Although some high tech workers share extremely high levels of education, training, and experience, another sizable part of the work force has virtually no documented skills or training. This two-tier pattern of skill level across the work force reflects the two-tier nature of much work in the industry.[10] Some work is highly professional—marked by demand for scientific and technical skills practiced more or less autonomously; other work is routine and repetitious, with minimal skill requirements.

This split has implications for human resource management, as Miljus and Smith (chapter 7) suggest when they allude to the impact of skill differentiation within the work force—for compensation policy, surely, but also for employee relations, recruitment, training, and worker productivity. These implications exist, in fact, in regard to all aspects of human resource management. They raise the issues of the degree to which particular human resource policies apply to disparate groups within the work force and of the integration of complementary or conflicting systems of human resource management within the firm.

Conclusions

In regard to human resource management in high tech, it is clear that two related phenomena are occurring. First, numerous high tech firms are indeed operating on the basis of innovative, well-integrated systems of human resource management. Many of their workers are a new breed, as the media suggest. They are professionals, and they are highly motivated. Their productivity levels are high. They are workaholics, and the ways in which their work is done, as well as the organization of the work place, are oriented to accommodate them—resulting in payoffs in product innovation and superiority. In total, the human resource management practices in such high tech firms present an identifiable new system that is innovative and effective.

At the same time, many high tech firms do not follow such successful human resource management models of flexibility and innovation. Some of these firms are new and subject to early mortality (Freeman et al., 1983); they are not yet in business for the long run. Such firms exist—consciously or not—in a temporarily sheltered niche of the market and take their profits as the opportunity allows. Their adoption of human resource management policies is ad hoc. Foulkes's case study of AutoTel, Inc. (chapter 5), suggests that, early on, many firms do not follow conscious or integrated strategies in regard to recruitment, promotion, or compensation. The perception of the need for and adoption of longer-term integrated policies evolves as—and if—the firm survives and grows. Many of the practices that the media identify as special to high tech have evolved in such firms. These firms may be especially vulnerable to piecemeal adoption of practices that have become socially legitimized (Tolbert and Zucker, 1983) but are not necessarily appropriate for the particular firm's needs at the time.

Some high tech firms are traditional companies that have moved into production of high tech products. The organization of such companies generally follows patterns set decades ago, when more hierarchical and centralized patterns were the norm (Stinchcombe, 1965). Although some new, more flexible practices may have evolved within the company, most human resource management policies follow along well-established lines.

Still others are high tech firms that have moved past the entrepreneurial start-up phase and whose product lines have matured and/or whose activities are concentrated in mass production. These firms continue to need to motivate and encourage professionally trained research and management personnel in order to remain competitive at the product innovation level, while their methods of mass production and assembly seem to demand and appropriately exercise economies in regard to both capital and labor investment. Such firms may be especially torn between the costs of maintaining, for all workers, practices that evolved in response to needs to recruit and ensure the loyalty and performance of highly skilled and educated workers and their vulnerability to charges of two-tier, seemingly elitist systems of work organization, employee relations, and compensation.

The questions then become: For which firms are the new human resource management practices functional? What are the possible modifications? What are the potential drawbacks of such practices? Finally, are issues emerging relative to high tech that will become salient for firms as the industry matures? The perspectives of this book speak to these questions and issues.

Notes

1. In geographic studies of the semiconductor industry in Orange County and Southeast Asia, Scott (1985a, 1985b) notes a diffusion of high tech industry over time, related to two major factors. The first, vertical disintegration—with various stages of design, development, production, and marketing subcontracted to branches or outside firms—is increasingly a factor in many manufacturing industries. Scott also notes a second factor: growth in the numbers of firms surrounding the typical branch plant or subcontractor, which provide it with upstream and downstream products as well as contracting-out services. Thus, in Southeast Asia, for example, there spring up near the typically low-skill electronic assembly plants other upstream firms that supply metal products needed in assembly. Other divisions or firms develop as well, to handle downstream testing stages and marketing. Craftsmen and technicians for these upstream and downstream firms are recruited from local labor markets, expanding the demand for skilled labor in those areas. Thus, the displacement of low-skilled jobs from the United States also leads to a complementary increased displacement of higher-skilled jobs.

2. Fein (1976), citing an extensive international literature on thirty years of worker participation programs in industry, comments: "Participation and involvement of employees in their work are among the most frequent suggestions by behaviorists and management theorists of ways to improve human performance and organization effectiveness" (p. 470). Fein notes that participation programs seem to have been most effective for white-collar jobs, and he comments, in regard to most routine jobs, that psychologists err when they propose that workers need to find fulfillment in work, that all work can be restructured so that it is fulfilling, and that enriched work will increase a worker's will to work (pp. 501, 523). Fein concludes that low-skilled work in itself is not demeaning, but lack of job security and low income are. Goals targeting increased productivity should therefore be linked to financial incentives, according to Fein. The nature of the job is also related to whether participation affects worker productivity. The reason for the absence of clear results from worker participation programs, suggests Derber (1970), is that much work does not demand autonomy or challenge. Although case studies such as that of Lincoln Electric (Fein, 1976) are often included in worker participation discussions, productivity in such firms is more clearly linked to financial incentives and to individual effort than to worker participation more broadly defined.

3. Not all education in the United States is oriented toward producing these worker characteristics. See, for example, Bowles and Gintis (1976), Oakes (1982), Giroux (1983), and Berg (1970), who note that schooling in the United States is variously targeted to socialize future workers into ways of working that will meet industry's needs. Thus, they argue, although some youth—especially middle-class and college-track youth—are socialized into professional attitudes toward work that encourage initiative, autonomy, and responsibility, many others receive schooling that prepares them for work that is routine and without autonomy. This conditions them to question ambiguity and to seek clarification from authority, which prepares them to tolerate boredom and also precludes the exercise of initiative.

4. For characteristics that define professionals and professional work, the discussions of Wilensky (1964), Kornhauser (1962), Strauss (1963), and Moore (1970) are especially relevant. Moore emphasizes the importance of education in the socialization of the professional. Education provides the technical language of the profession, which is not only the code by which professionals recognize each other but also the credential—the tangible sign that the code has been taught. According to Wilensky, much of the knowledge that defines a profession is tacit and relatively inaccessible unless it is called forth by a particular task. Technical language, whether it is the jargon of computerese or the *Physician's Desk Reference* or the French of Cordon Bleu, assures that the professional knows what he is talking about—and that he is solidly grounded in that expertise. The manual skill by which one craftsman acknowledges another's work and thus accepts evaluation has a modern-day parallel in the scientific professionals' shared technical language, which is relevant to evaluation and supervision, as discussed later.

5. In chapter 4, Pister discusses the 125 percent increase of professionally educated engineers in industry since 1970. Similar increases may be traced in the various U.S. Censuses of Population, 1950 to 1980.

6. See, for example, Kornhauser (1962) and Strauss and Rainwater (1962).

7. Results of a 1985 UCLA survey of top-level managers of companies in transition (conducted by the editors in cooperation with the American Electronics Association) showed unionization at the bottom of the list of salient human resource management issues.

8. The widespread increase in unionization of professional workers in the public sector contrasts with this experience. Today, unionism in the public sector is on the increase—a clear-cut reversal of the trend toward decreasing unionization in the private sector. Acceptance of unionization by professional workers is said to have influenced this trend. Estimates of the extent of collective bargaining organization among public school teachers, for example, is estimated to be 75 percent nationwide (Bureau of National Affairs, 1980, 412). Overall, unionization of professional workers is much higher (29.7 percent) than unionization among other white-collar workers, even in the private sector (Dodd, 1979).

9. The evolution of the role of the research administrator is documented by the analyses of Kornhauser (1962) and Strauss (1963), who indicate the need for joint supervision by both expert and manager; by Wilensky (1964), who notes that if the industrial lab were run by a scientist-administrator, it would facilitate administrative policies aimed at preserving professional autonomy; and in case studies such as that of the Rand Corporation (Smith, 1966), in which it is clear that the development of the matrix organization accommodated the evolution of the role.

10. Suggestions that high tech is characterized by a two-tier wage structure, with a large group of educated, skilled, and high-paid scientists, technicians, and managers on the one hand and another group of virtually unskilled and very low-paid production workers on the other are rampant in the critical literature. This is another, perhaps crucial, aspect of the paradox of high tech. Does high technology industry contribute to increasing income inequality in the United States? This question was assessed by Medoff and Strassman (1985), who checked the relationship between high tech and two-tier distributions of wages and skill levels. Using company wage data, Medoff and Strassman show that no two-tier patterns exist. They suggest that this is not because of obvious and

acknowledged differences in pay and skill between highly skilled, highly paid scientists who design and develop the products and the low-skilled, low-paid workers who assemble the parts. There are marked wage disparities here, and they can be linked to skill level. However, the manufacture of high tech products has a job-multiplier effect, creating and sustaining jobs in the firm's sales, communications, and other white-collar areas. These middle-level jobs fill in the gap between the highly skilled, highly paid, professionally trained workers and the low-skilled, low-paid workers. Or many companies may simply export wage bimodality when components are produced by low-skilled, low-paid workers employed in branch plants overseas, prior to testing and sales in the United States. Medoff and Strassman note that in their sample, there was more wage disparity in companies that were less vertically integrated, which also suggests that a composite picture of the work force necessary to design, produce, sell, and service a product would reveal a less bimodal distribution of wages.

Part II
The High Technology
Work Force

The theme that ties together the three chapters in this part is a concern with the quantitative and qualitative adequacy of the work force needed to support high technology development in the United States over the next generation.

Each of the authors assumes that universities and other educational institutions play a crucial role in maintaining the vitality of the important high technology sector, including training the work force of engineers, technicians, computer scientists, and biologists.

Although the authors focus on different aspects of this development, the chapters reinforce each other in that they put in perspective the key policy and operational issues for industry, educational institutions, and government that need to be dealt with to ensure adequate human resources for high technology industries. Because of the dependence of these industries on research and product innovation, the authors deal principally with questions related to engineers, scientists, and other highly skilled professionals.

The central issue of the projected demand for and supply of innovative and highly skilled scientific and technical workers is addressed by Richard Belous in chapter 2. Forecasting human resource requirements for high technology industry presents formidable definitional and methodological problems, and Belous acknowledges these, using alternative definitions of high technology industry with a methodology developed by Leontief and Duchin (1985). Belous considers the overall demand for scientific personnel and finds that even if high technology is defined very broadly, by 1995 it will still claim a relatively small share of the total industrial demand for skilled workers. He finds little evidence for the proposition that there will be a generalized shortage of technical personnel to serve the needs of high technology, and he questions the necessity of new, large-scale training programs to meet future demand. At the same time, Belous examines the extent to which the fairly optimistic situation depicted by the overall, aggregate analyses may mask local and temporal shortages and problems in regard to an adequate work force.

Human resource managers in high technology firms consistently identify recruitment and development of their highly skilled scientific and professional work force as top-priority activities. The adequacy of supply and the education of this crucial segment of the high technology work force are important issues for government policymakers and educational institutions.

In chapter 3, Lewis Solmon and Midge La Porte focus on the quality of the work force being produced for high technology industry, an issue that Belous would agree is more urgent than numerical adequacy. Solmon and La Porte examine a variety of data to review the contributions of different educational institutions in the preparation and maintenance of an adequate labor supply. They identify trends that suggest that the quality of the pool available to be trained in the United States may be contracting on both an absolute level and also relative to increasing supplies of highly trained scientists and engineers abroad. The authors also examine the hypothesis that our primary and secondary schools are doing an inadequate job of instilling the basics, which severely limits the quality of graduates that eventually are produced for the job market.

Solmon and La Porte devote much attention to the role of education and training carried out by high technology industry itself. They also address issues related to predicted shortages of teachers and professors of future scientists and engineers and the ways in which high tech industries may be inadvertently restricting their own future supplies of such workers by hiring the most talented away from academe.

Both the Belous and the Solmon and La Porte chapters mention the fact that an increasing share of the basic research that drives the engine of high technology development and that helps to maintain a strong position vis-à-vis international competition is carried on within industry. This represents a shift away from universities in terms of the relative share of research and development dollars historically devoted to university-based research.

In chapter 4, Karl Pister discusses the implications of these and related developments for both the university community and industry. He emphasizes the inescapable interdependence of educational institutions (especially research universities), high technology industry, and government in ensuring that both the research and the quality of training needed to keep the United States at the cutting edge of technical innovation can continue apace. Pister also describes some exemplary programs that link industry with academe and discusses their advantages as well as some possible disadvantages and risks.

2
High Technology Labor Markets: Projections and Policy Implications

Richard S. Belous

igh technology is like pornography—we all think that we know what it is, yet few can adequately define it. Intuitively, we may think of images of silicon chips or white-coated scientists in germ-free environments, splicing genes. Yet our actual basic definitions of high tech are a long way from our intuitive images. Specific definitions seem to leave out many people and companies that we tend to think of as part of this sector. However, broader definitions often wind up including the ridiculous, as well as the sublime, in high tech.

Problems of definition have not stopped congressional leaders from forming plans that would "encourage the most dynamic sector of our society" (*Congressional Record*, 1985, S431). Neither have they daunted the zeal of many analysts. For example, one such analyst recently warned that "the real measure of success between competing knowledge intensive economies will be found in the quality of their human resources" (Botkin, Dimancescu, and Stata, 1982, 47). Yet basic tools of labor market analysis are often not employed in these types of studies. Clarion calls seem to be easier to make than actual attempts to measure "knowledge intensity" and "quality."

The bottom line of this chapter is that many U.S. high technology industries and labor markets face serious problems—but they are **not** the types of problems that are often envisioned by "future shock" analysts. These problems would not be solved if the United States doubled its number of high technology scientists

I wish to thank Wassily Leontief, Faye Duchin, and their colleagues at New York University's Institute for Economic Analysis for all their help. This chapter was improved by advice from Carolyn Kay Bracato, Mary Jane Bolle, Gary Guenther, and John Williamson, of the Congressional Research Service's Economics Division. I also wish to thank Sar Levitan, of George Washington University, for his comments. Only wisdom—no mistakes—should be ascribed to the aforenamed. This chapter represents my personal opinions and does not necessarily represent the opinion of The Conference Board, the Library of Congress, or the aforenamed individuals.

and engineers or if every sixth-grader could run regressions on his or her personal computer. Instead, the problems often center on the more mundane micro-economic aspects of labor markets (for example, institutional constraints, lagged responses, and poor information). The real difficulty in coping with U.S. high technology labor problems may be that what are called for are **not** sexy solutions, yet half the fun and promise of high technology (at least from the policymakers' point of view) is its sex appeal.

To demonstrate these points, the next section of this chapter will use a dynamic input/output model constructed by Wassily Leontief and Faye Duchin to examine the terrain of U.S. high technology labor markets under various assumptions and definitions. The following section will then explore how the U.S. high tech sector stands in the face of stiff international competition, and the final section will focus on some of the policy implications of these findings.

The U.S. High Technology Employment Scene

Although there is no widely accepted definition of high technology industry, researchers have tended to use four types of factors in examining high tech employment. First, they have considered the utilization of "technology-oriented" workers (often defined as engineers, life and physical scientists, and technicians in those fields). Second, they have looked at an industry's expenditures on research and development. Third, they have tried to make a qualitative judgment about the nature of the goods or services produced by an industry. And fourth, they have tried to make qualitative judgments concerning the actual technology employed or being developed by an industry (Riche, Hecker, and Burgan, 1983, 50–53; Dorfman, 1982, 17–20).

If an industry equals or is above some threshold in terms of technology-oriented workers, R&D expenditures, or qualitative judgments concerning products and technology, then that industry is counted as part of the high tech sector of the economy. High tech employment levels would then be the summation of the human resource requirements of the various industries included. High tech industries might be defined on the basis of any one of the factors cited or on some combination of these variables (Breckenridge, 1983, 1–6).

As indicated in table 2–1, a very broad definition of high tech industries would indicate that more than 13 percent of the U.S. work force is employed by this sector. However, this definition (which is based only on the relative number of technology-oriented workers in an industry) includes such industries as farm and garden machinery, soaps, cleaners, and toilet preparations. A very narrow definition of high tech indicates that less than 3 percent of the U.S. work force is employed by this sector. However, this narrow definition (which is based only on

Table 2–1
U.S. Employment in High Technology Industries under Various Definitions

	Percentage of the U.S. Labor Force		
Definition of High Tech	*1959*	*1982*	*1995[a]*
Very broad[b]	13.1	13.4	14.1
Very narrow[c]	2.7	2.8	2.9

Source: Riche et al., 1983, 50–53.

[a] Estimates are based on the moderate growth projection of the U.S. Bureau of Labor Statistics input/output model of the U.S. economy.

[b] The very broad definition counts an industry as part of the high tech sector if its relative employment of technology-oriented workers equals 150 percent of the rate of all U.S. industries.

[c] The very narrow definition counts an industry as part of the high tech sector if its ratio of R&D expenditures to sales is at least twice the average for all industries.

relative R&D expenditures) excludes noncommercial educational, scientific, and research organizations.

Although the level of relative U.S. high tech employment is quite different depending upon the definition of high tech industry, table 2–1 does indicate several general trends. First, under both definitions, high tech employment has been growing and will continue to grow at a slightly faster pace than employment rates for the entire U.S. economy. Second, despite this slightly faster employment growth rate, only a small minority of new jobs will be in the high tech sector. Under the very broad definition, only about 17 percent of the new jobs created in the 1982–95 period will be in the high tech sector (Riche, Hecker, and Burgan, 1983).

Thus, even under very broad definitions, high tech will be able to absorb only a small fraction of the growing U.S. labor force. However, although high tech's slice of the employment pie may be small in quantitative terms, there are reasons to believe that it could be very significant in qualitative terms. In the past, analysts have often pointed to the small relative employment levels and share of total gross national product represented by such industries as steel or railroads. Yet despite their quantitative employment and output levels, these industries have had a significant impact on labor and product market patterns in the U.S. economy. High tech labor markets could also have an influence beyond their relative size. For example, other industries might attempt to emulate the way labor–management conflicts are resolved in high tech industries. Also, the health of many non-high tech U.S. industries could be influenced by the relative strengths of U.S. high tech firms. Hence, it is important to consider the indirect as well as the direct effects high tech labor markets can have on the entire U.S. economy.

The Leontief-Duchin Model

Given the current state of the art, the Leontief-Duchin dynamic input/output (I/O) model of the U.S. economy is probably the best analytical tool for capturing some of the indirect as well as direct effects noted above. The Leontief-Duchin model not only considers the final products produced by the U.S. economy but also examines the impact of shifting technology on the numerous intermediate stages of production. It is dynamic in the sense that it provides employment (by industry and occupation), output, and investment time paths over many years. It is also a dynamic model in the sense that it allows the impact of shifting technology to filter through all of the various industries and occupations in the United States. Reactions in one industry and labor market can have a ripple effect that influences other sectors of the economy. Eventually, this chain of events can have a feedback effect on the original industry or occupation that started the ripple.[1]

Since a good deal of the analysis in this chapter is based on the Leontief-Duchin model, it is important to explain the essence of this analytical tool. At the heart of the Leontief-Duchin I/O model are four types of tables, or matrices. A first matrix represents the input requirements of each industry to meet current consumption demand, and a second matrix is designed to capture capital expansion of the various industries. A third matrix is built to capture the capital replacement requirements of each industry, and the diverse labor inputs needed by various industries are modeled in a fourth matrix. Finally, a series of columns, or vectors, is designed to capture noninvestment final demand, including household consumption, government purchases, and exports and imports. (High tech export and import trends will be examined in the next section.)

In this model, the U.S. economy is disaggregated into eighty-nine industries and fifty-three occupations. All of these matrices and vectors can be expected to shift over time. One of the major factors behind the need to alter the matrices is, of course, the introduction and diffusion of advanced technology. The adjustment of each matrix and vector has required sector studies. For those analysts who do not like computer models, the Leontief-Duchin research effort has something to warm the heart of even the most traditional institutional economist—Leontief, Duchin, and Szyld (1985) and their colleagues (see note 1) conducted a good deal of direct research on such new industries as robotics and computer-based instruction.

Of course, estimates produced by this or any model should be considered, at best, as rough approximations, not as exact numbers. Nevertheless, I believe that this model can provide new insights into the nature of U.S. high tech labor markets. Some have faulted I/O models because of their so-called rigidities (that is, lack of full price and income substitution effects). However, these complaints often ignore or show an ignorance of I/O model advances of the past ten years. Also, some have suggested that general equilibrium models should be used instead of I/O analysis. These complaints ignore reality. I hope that in the future, general equilibrium models will have advanced to the point that they can be used to study the questions discussed in this chapter; however, they are **not** at that point today.[2] For these reasons, I believe that it is wise to use I/O analysis despite its imperfections.

High Tech Patterns

The Leontief-Duchin model was combined with a slightly modified definition of high tech. In this case, high tech was considered to be all industries in which (1) the ratio of R&D expenditures to sales was at least twice the average for all industries and/or (2) the relative employment of technology-oriented workers equaled 300 percent of the rate of all U.S. industries. This is a mildly conservative definition of high tech; under this definition, the following industries can be considered high tech: chemicals, drugs, electronic computing equipment, semiconductors and related devices, electronic components, aircraft and parts, communications, robotics, and computer-based instruction.

As indicated in table 2–2, about 4.6 percent of U.S. employment is currently in high tech industries (as defined here), and employment in this sector should grow to about 5.3 percent of the U.S. total in 1990. So far, the general trends presented in table 2–2 conform with the general trends outlined in table 2–1. Although high tech employment should be growing at a faster pace than the entire economy, only a small minority of jobs will be in this sector. Under this mildly conservative high tech definition, only about one out of seven new jobs will be in these industries during the remainder of the 1980s. Thus, these findings (concerning growth rates and share of new jobs) seem to be very robust under various high tech definitions.

Table 2–2 indicates several key occupations in high tech industries. Although the high tech sector represents only about one out of twenty-three U.S. jobs, it accounts for almost three in eight electrical engineering jobs, one in five industrial engineering jobs, and one in six mechanical engineering jobs. More than 17 percent of natural scientists are employed in high tech, and more than 20 percent of computer system analysts work in this sector.

It is not unexpected that high tech should represent a significant chunk of the U.S. engineering and scientific work force, but table 2–2 also indicates other important labor market factors. For example, high tech's share of U.S. employment in several computer-related occupations should decline somewhat as the microelectronic revolution spreads to other parts of the economy. As indicated in figure 2–1, the relative size of the white-collar labor force in high tech is larger than that in several traditional "smokestack" industries. Yet many high tech industries still include a significant blue-collar work force. In electronic components, for example, 61 percent of the work force is employed in blue-collar jobs;

Table 2–2

Employment Levels, Growth Rates, and High Tech's Share of U.S. Civilian Employment

Occupation	Employment in All High Technology Industries[a] 1985	Employment in All High Technology Industries[a] 1990	Growth between 1985 and 1990 (%)	High Tech's Share of U.S. Employment (%) 1985	High Tech's Share of U.S. Employment (%) 1990
Total high tech employment	4,741,394	5,842,926	23.2	4.6	5.3
Electrical engineers	139,648	192,616	37.9	35.9	38.3
Industrial engineers	44,646	58,468	31.0	20.7	22.9
Mechanical engineers	37,874	46,770	23.5	16.1	17.5
Other engineers	102,018	115,231	13.0	18.6	19.0
Natural scientists	56,586	67,027	18.5	17.2	18.0
Computer systems analysts	59,522	80,548	35.3	20.4	18.0
Computer programmers	72,345	99,176	37.1	15.0	13.2
Managers and proprietors	383,606	453,845	18.3	3.6	4.1
Sales workers	96,524	113,708	17.8	1.4	1.5
Secretaries	212,309	211,426	−0.4	4.7	5.5
Machinists	59,050	67,552	14.4	11.1	12.4
Tool and die makers	23,106	26,556	14.9	11.7	14.2
Assemblers	342,435	477,937	39.6	23.5	27.2
Welders	43,284	59,068	36.5	5.7	7.4
Protective service	19,268	24,037	24.8	3.1	3.4
Laborers	88,441	117,091	32.4	1.6	2.0
Janitors and sextons	44,358	58,131	31.0	2.8	3.3

Source: The Leontief-Duchin model.

[a] High technology industries are defined as industries in which (1) the ratio of R&D expenditure to sales was at least twice the average of all industries and/or (2) the relative employment of technology-oriented workers was 300 percent of the rate of all U.S. industries (see Riche et al., 1983, 50–53; Dorfman, 1982).

and almost 38 percent of the robotic industry's work force is employed in blue-collar jobs.

Although high tech may represent about 5 percent of all U.S. employment, high tech's current share of all U.S. machinists is over 11 percent (see table 2–2). Also, high tech's share of all U.S. tool and die makers should be over 14 percent by 1990. At the same time, more than one in four U.S. assembly workers should be employed in the high tech sector by 1990. In contrast, only about 1.6 per cent of all U.S. laborers are employed in high tech industries, but this share seems to be growing in both absolute and relative terms.

The differences between high tech human resource requirements and the human resource requirements for all industries are made clearer by looking at

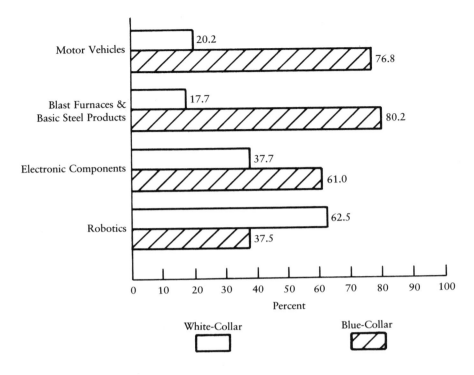

Source: The Leontief-Duchin model, and U.S. Bureau of Labor Statistics.

Figure 2–1. The Relative Size of White-Collar and Blue-Collar Employment in Various Industries, 1980

relative occupational employment. As shown in table 2–3, engineers represent about 1.3 percent of employment in all U.S. industries. However, engineers represent 6.8 percent of employment for high tech industries. Although less than 1 percent of employment in all industries is in direct computer occupations (systems analysts, programmers, and other computer specialists), 3 percent of high tech workers are employed in these occupations. In contrast, the relative occupational composition of the high tech sector's work force is quite similar to that of the total economy for some classifications, such as secretaries and other clerical workers.

Beyond differences in white-collar work forces, the high tech sector also shows significant differences in its blue-collar work force when matched with the blue-collar work force in many other industries. For example, whereas assemblers are only 1.4 percent of all industry employment, they constitute 7.2 percent of high tech employment. In fact, for the robotics and semiconductor industries, assemblers represent more than 15 percent of the work force. Although the focus

Table 2-3
Relative Occupational Employment in Various High Technology Industries and the Total U.S. Economy, 1990
(percentages)

| Occupation | High Tech Industry | | | | | | All High Tech Industries[a] | Total U.S. Economy |
	Chemicals	Drugs	Electronic Computing	Semiconductors	Aircraft	Robotics		
Engineers	5.2	2.0	9.4	6.6	14.7	28.4	6.8	1.3
Electrical	0.4	0.1	4.2	4.0	2.2	12.8	2.9	0.4
Industrial	0.5	0.5	1.8	1.2	1.7	2.3	0.9	0.2
Mechanical	0.6	0.3	1.7	0.6	2.0	4.6	0.8	0.2
Others	3.7	1.1	1.7	0.8	8.8	8.7	2.2	0.5
Health technologists	0.1	0.4	0.0	0.0	0.0	0.0	0.0	1.1
Physicians, surgeons	0.0	0.1	0.0	0.0	0.0	0.0	0.0	0.4
Others	0.1	0.3	0.0	0.0	0.0	0.0	0.0	0.7
Computer workers	0.6	0.8	13.3	1.1	1.8	3.4	3.0	0.9
Systems analysts	0.2	0.4	5.8	0.4	0.6	0.9	1.3	0.3
Other computer specialists	0.0	0.0	1.1	0.1	0.2	0.3	0.2	0.1
Programmers	0.4	0.4	6.4	0.6	1.0	2.2	1.5	0.5
Natural scientists	4.0	5.5	0.5	0.3	0.6	0.0	1.2	0.3
Other professional and technical workers	10.4	10.2	13.6	6.8	9.5	8.6	8.1	6.1
Managers, proprietors	8.5	10.2	9.6	6.2	6.5	8.8	8.1	10.4
Secretaries	4.1	6.8	5.4	3.7	3.9	4.8	4.5	4.2
Other clerical	6.4	8.6	7.1	6.9	7.8	8.1	10.4	8.5
Machinists	1.4	0.7	0.9	1.7	3.4	2.2	1.2	0.5
Tool and die makers	0.1	0.1	0.4	0.8	1.6	0.0	0.5	0.2
Assemblers	0.9	3.7	9.6	15.9	7.6	15.2	7.2	1.4
Laborers	4.4	3.6	0.7	2.6	0.9	0.7	1.9	5.2
Other occupations	53.9	47.4	29.5	47.4	41.7	19.8	47.1	59.9
Total %	100.0	100.0	100.0	100.0	100.0	100.0	100.0	100.0

Source: The Leontief-Duchin model.
Note: See table 2–2, note a, for the definition of high technology.

is often on high tech white-collar workers, the shape of the high tech blue-collar work force could have a significant impact on international trade issues and, of course, on labor–management relations.

Supply and Demand

A major concern expressed by many analysts and policymakers is whether or not the United States will be able to meet its high technology human resource requirements. Will the demand for highly educated and highly trained technology-oriented workers be far greater than the supply? Should the United States vastly expand the resources allocated to the education and training of these types of workers? Some have noted that Japan, for example, has been graduating a far larger number of engineers than the United States. Might this condition come back to haunt us, as an excess demand for engineers and scientists constrains the rate of potential U.S. growth? This imbalance in supply and demand of key high tech people could create a "new scarcity," according to some (Botkin et al., 1982, 49–70; National Academy of Sciences, 1983a, 46–50).

There are indications that the demand for many types of technology-oriented workers will continue to grow at a much faster pace than the average rate of job growth for the entire U.S. economy (see figure 2–2). For the rest of this century, the employment growth rate of electrical engineers could be twice the rate of total employment. At the same time, the employment growth rate of computer systems analysts could be more than four times the rate of the total labor force.

Nevertheless, a strong growth in demand does not mean that supply-side forces will be inadequate to meet these requirements. Although our abilities to model the supply side have often been behind our demand-side models, a recent effort by Robert Dauffenbach and Jack Fiorito (1983) is significant. The Dauffenbach-Fiorito work is a stock/flow supply model that attempts to explain and predict the various components of the labor supply of scientists and engineers. Given an existing stock, the model explores the net supply of scientists and engineers as individuals "flow" (despite many institutional rigidities) into various parts of the economy. The supply of science and engineering personnel is seen as consisting of four major segments: (1) the initial stock of scientific and engineering personnel; (2) the flows of new entrants into these occupations; (3) the flows of experienced workers into, within, and out of various scientific occupations; and (4) international flows of scientific and engineering workers. Each segment of supply is viewed as responding to various economic (including demand-side) forces and institutional variables. The demand side of the labor markets for scientists and engineers can be estimated by using the input/output concepts described earlier (National Science Foundation, 1984). Since demand and supply conditions are interrelated, the demand-side and supply-side models

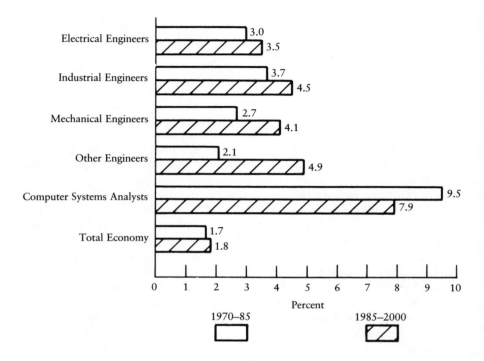

Source: The Leontief-Duchin model, and U.S. Bureau of Labor Statistics.

Figure 2–2. Average Annual Rate of Employment Growth for Various Occupations

should not be run totally independent of each other. Also, one can make any number of assumptions to produce various scenarios.

Table 2–4 is based on high real macroeconomic growth assumptions and high real defense spending assumptions. The reasons for picking these assumptions is that if a scarcity of scientists and engineers were to materialize, it is reasonable to expect that it would be under these conditions. However, scarcity (or excess demand) appears in only a few occupations (aeronautical/astronautical engineers, industrial engineers, and computer specialists). Again, the table 2–4 estimates are based on very high (and what to many seem unrealistic) macroeconomic growth and defense spending assumptions; thus, the table might be considered an extreme case (very tight labor market). Yet even in this extreme case, the supply and demand conditions for many types of engineers and scientists do **not** show a dramatic imbalance.

Table 2–4
Projected Supply/Demand Balance of Scientists and Engineers under High
Growth and High Defense Spending Conditions, 1987

	Excess Supply (thousands)	Excess Supply as a Percentage of Supply
Total scientists	+ 55.2	8.6
Agricultural	+ 3.8	17.1
Biologists	+ 13.8	18.1
Chemists	+ 1.8	1.7
Geologists	+ 3.7	7.2
Mathematicians	+ 2.4	3.8
Physicists	+ 2.4	9.0
Other life and physical scientists	+ 2.0	6.0
Social scientists	+ 25.3	9.6
Total engineers	+ 14.2	1.0
Aeronautical/astronautical	− 4.4	4.2
Chemical	+ 3.3	5.1
Civil	+ 1.2	0.6
Electrical/electronic	+ 0.7	0.2
Industrial	− 1.0	0.8
Mechanical	+ 5.2	2.1
Metallurgical/mining	+ 0.7	3.6
Petroleum	+ 0.9	2.8
Other engineers	+ 7.6	3.4
Computer specialists[a]	−11.0	1.9

Source: Dauffenbach and Fiorito (1983); and National Science Foundation (1984).
Note: The demand estimate is based on high macroeconomic growth assumptions (that is, GNP increases 4 percent per year) and high defense spending assumptions (that is, between 1982 and 1987, Defense Department real spending increases by 45 percent). A plus sign (+) means that supply is greater than demand; a minus sign (−) means that demand is greater than supply.
[a] Includes computer systems analysts and programmers.

Instability

Other factors should be considered besides the balance of long-term supply and demand forces. Although there may be a degree of total long-term equilibrium, there could be a significant short-term instability in many of these labor markets. In the last recession, for example, high tech industries showed relative employment declines that were larger than the total U.S. nonfarm economy. However, relative employment declines for high tech were not as large as those for all manufacturing. High tech is not immune to the business cycle (Burgan, 1985).

Many scientific and engineering labor markets appear to be classic cases of what economists call "cobweb" cycles. These cycles happen when the supply side responds only very slowly to the demand-side forces of the market. For example, as demand for various types of scientists and engineers shifts, the supply response is delayed. It would be as if the net supply in 1985 of "new" scientists were the result of economic and social forces that existed in, say, the 1975–83 period. Meanwhile, the demand for these "new" scientists in 1985 might hinge on many expected future events (for example, congressional action on 1986 defense spending). The net result could be compared to a pack of race dogs following a mechanical hare, except that in this analogy—unlike a real dog race—the movements of the mechanical hare (demand) are often highly erratic (Freeman, 1976; Watson, 1972, 279–84).

Such a "cobweb" cycle can produce a good deal of short-term instability, even if the long-term supply and demand forces are in balance. Figure 2–3 presents just this situation. The projected time path for the employment of electrical engineers indicates a very bumpy future—despite significant long-term growth—under each of the scenarios presented in the figure. This short-term erratic behavior can hurt the U.S. future in high tech. But this problem (and the potential solution) is quite different from fears of long-term shortages. Even if we matched the Japanese in the exact number of engineers graduated from U.S. universities each year, this short-term erratic behavior would remain. As indicated later in this chapter, a good solution might lack the "sex appeal" of programs designed, say, to double the number of U.S. engineering graduates over current levels.

Eating the "Seed Corn"

The biomedical sciences present a good case study of other types of difficulties in scientific and engineering labor markets. As indicated in table 2–4, there does not appear to be a shortage of biologists. However, there may be some problems in the allocation of biomedical scientists among various types of employers. Prior to the recent growth of several biomedical industries and new production technologies, the general career path in this field included a university doctoral program. With the expansion of the number of college students in the 1960s and early 1970s (due in large part to the "baby boom" generation), the academic demand for biomedical scientists with doctorates was strong. Also, increased government research funding boosted the number of postdoctoral appointments—that is, appointments that were not part of the normal university tenure-track system (National Academy of Sciences, 1983b, 50–87; Normin, 1981, 121–29).

However, many of these forces started to move in the opposite direction in the late 1970s and early 1980s, and the demand for biomedical scientists by the academic sector slowed down—partly because the last of the "baby boom" generation was entering college and partly because of some reductions in federal

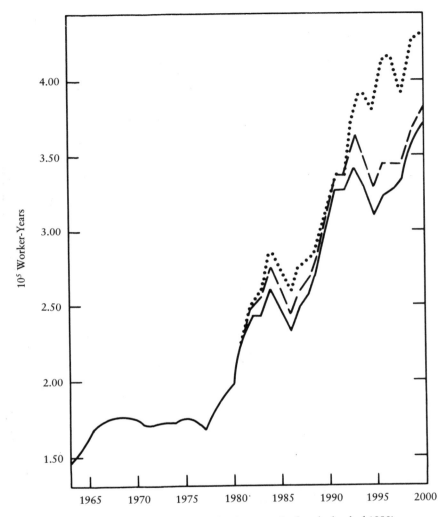

Scenario 1: Baseline (i.e., if future technology remained at the level of 1980).
Scenario 2: Moderate (i.e., if introduction of future technology is at a moderate pace).
Scenario 3: Rapid (i.e., if introduction of future technology is at a rapid pace).

Source: The Leontief-Duchin model, and U.S. Bureau of Labor Statistics.

Figure 2–3. Estimated Employment of Electrical Engineers in the United States

funding. At the same time, private industry has provided a growing and lucrative alternative to a doctoral program in the biomedical field. With the prospects of much greater financial rewards for working in private industry than for working in traditional academic and research areas, many bright minds have given up on the halls of higher learning. Or as one National Academy of Science (1983b) panel put it:

> Bioscience Ph.D. production has been essentially level since 1972 and can be expected to start dropping. . . . Young researchers will find it difficult in these circumstances to begin their careers as independent investigators. Without adequate numbers of young investigators, who typically are highly innovative and creative, where will the new ideas for advances in basic research come from? . . . How can this country's competitive advantage in technological areas, such as the new biotechnology, be maintained? (pp. 6, 50)

The argument is that this reallocation of bioscience workers (away from more traditional academic and research areas) will, in effect, eat up America's vital "seed corn" of human capital. Yet, like the "cobweb" cycles discussed earlier, these labor allocation problems are more microeconomic in nature than macroeconomic, and macroeconomic solutions may be of little use.

The International Scene

Faced with a large trade deficit and several limping "smokestack" industries, some policymakers and analysts have looked toward the U.S. high tech sector as the great hope for America's future. Although the United States may have problems with traditional export areas (for example, steel, basic machinery, and the like), high tech exports could take up some of this slack, it is argued.

However, it appears that the United States is going to face very stiff international competition in the high tech field. As indicated in table 2–5, high tech's relative share of total U.S. exports was 12.3 percent in 1977. By 1990, (according to Leontief-Duchin model projections), high tech's relative share of total U.S. exports is projected to increase very little, to 13.9 percent. At the same time, high tech's share of total U.S. imports was 4.4 percent in 1977. But by 1990, high tech's relative share of total U.S. imports could grow to 6.4 percent. To keep these percentages in perspective, in both 1977 and 1990 (estimated), the relative share of agricultural exports to total U.S. exports is 14.3 percent. Thus, U.S. high tech exports could be showing only very modest relative growth. In many high tech areas, imports could capture a major (and growing) share of the U.S. market. For example, between 1977 and 1990, imports of computers, semiconductors, and aircraft should be making substantial gains in the U.S. market.[3]

Table 2–5
U.S. Exports and Imports in Various High Technology Industries
(millions of constant 1979 dollars)

A. Exports and Imports	Exports		Imports		Percentage Change 1977–90	
Industry	1977	1990	1977	1990	Exports	Imports
Chemicals	5,737	8,489	3,996	5,625	48.0	40.8
Drugs	2,161	4,337	897	1,263	100.7	40.8
Computers	2,689	7,295	939	2,730	171.3	190.7
Semiconductors	1,339	3,175	1,305	3,530	137.1	170.5
Aircraft	8,055	13,852	989	2,995	72.0	202.8
Robotics	N.A.	48	N.A.	238	N.A.	N.A.
Total high tech	19,981	37,196	8,126	16,281	86.2	100.4

B. High Tech Net Exports[a]
1977	$11,855
1990	$20,915

C. High Tech's Share of Total U.S. Exports
1977	12.3%
1990	13.9%

D. High Tech's Share of Total U.S. Imports
1977	4.4%
1990	6.4%

Source: The Leontief-Duchin model.
Note: See table 2–2, note a, for the definition of high technology.
[a] Exports minus imports.

The Engine of Growth

As the U.S. International Trade Administration recently noted:

> The U.S. competitive position deteriorated some years ago in the largest high tech category—communications equipment and electronic components. . . . There are some indications that the U.S. competitive position may be eroding in other high tech areas as well. . . . It is difficult to assess the extent to which the recent decline in U.S. high tech trade performance stems from short-term factors such as early U.S. economic recovery or an over-valued dollar, or from long-term factors such as a narrowing of the underlying U.S. technological advantage. (U.S. Department of Commerce, 1984, 15)

Many different types of economic and social factors determine a nation's comparative advantage in international markets. Yet, along with exchange rates, wage-related policies, which can range from statutory compensation levels to collective bargaining regulations and other labor laws, "are probably the most important determinants of a country's international competitiveness in the near term" (Koptis, 1982, 463). Thus, recent compensation trends in several key nations could have a significant impact on high tech trade flows.

In general, the gap between high U.S. wage rates and lower foreign wage rates paid to similar workers narrowed during most of the post–World War II era. However, in the early 1980s, this gap appeared to widen. Some policymakers have warned that this international wage gap could encourage the purchase of more foreign high tech items in the U.S. market and could encourage some corporations to build or relocate their high tech production facilities in foreign low-wage labor markets.

As an example of these trends, French manufacturing workers, on average, were paid 31 percent of the hourly compensation received by similar U.S. workers in 1960. The size of this gap was reduced between 1960 and 1980 (that is, average French manufacturing compensation equaled 92 percent of the level paid similar U.S. workers in 1980). However, between 1980 and 1983, the gap increased in size. French manufacturing workers, on average, were paid only 62 percent of the hourly compensation received by similar U.S. workers in 1983.

Despite all this, it appears that the primary factor behind the recent growth in the U.S.-foreign wage gap has **not** been labor market forces. Rather, shifts in foreign exchange rates, including a very strong U.S. dollar, seem to have played the leading role in this process. When foreign wage rates are stated in terms of U.S. dollars, the foreign wage rates will appear to decline as the value of the U.S. dollar rises.

Japan provides an example of the influence of foreign exchange rates on wage levels. In 1981, Japanese average wages paid in manufacturing were equal to 56 percent of the level paid similar workers in the United States. However, by 1983, Japanese average wages paid in manufacturing were equal to 51 percent of the level paid similar workers in the United States. If the U.S. dollar per Japanese yen exchange rate had remained constant in this period, Japanese average wages in manufacturing in 1983 *might* have been equal to 55 percent of the level paid similar workers in the United States. Thus, of the total five percentage point fall in average Japanese wages compared with average U.S. wages between 1981 and 1983, four percentage points appear to be due to foreign exchange rate factors and only one percentage point appears to be due to labor market forces (Belous, 1984b).

In recent years, foreign exchange rates have been influenced much more by large international capital flows and interest rates than by labor market conditions. Also, compensation differentials between the United States and some of its major trading partners may narrow in the future if the value of the dollar declines relative to other major currencies. Meanwhile, average wages paid in Third World developing nations have often been less than 15 percent of the level paid similar workers in the United States. Continued population growth, resulting in significant increases in Third World labor supplies, could create major downward pressures on many wage rates paid in Third World labor markets.

Sound international wage estimates disaggregated down to the industry level remain scarce. However, as shown in table 2–6, compensation estimates for production workers in the chemical, electric and electronic equipment, and aircraft industries indicate that international high tech compensation levels follow the general patterns described here.

International Competition

Low wage rates do not necessarily result in low unit labor costs (that is, the labor costs to produce a specific product). It is possible for a country to enjoy higher wage rates and productivity levels than other nations and, at the same time, experience lower unit labor costs than other countries.

Given the relationships between compensation, productivity, and unit labor costs, good estimates of these three variables could provide an analyst with many insights. For example, international trade flows of high tech products and the international location of high tech plants should be significantly influenced by

Table 2–6
Hourly Compensation for Production Workers in Various Industries in Various Countries, 1982
(index: United States = 100)

Country	All Manufacturing	Chemicals	Electric and Electronic Equipment	Aircraft
Israel	38	41	49	N.A.
Japan	49	61	46	N.A.
Korea	10	13	11	N.A.
Taiwan	13	13	12	16
West Germany	89	83	87	72
United Kingdom	58	57	57	52
Sweden	86	74	85	N.A.

Source: U.S. Department of Labor, Bureau of Labor Statistics, unpublished data.

these factors. Unfortunately, we remain in the "dark ages" about international productivity and unit labor cost estimates; sound data most often do not exist. Even on the domestic front the situation is bleak. Computers are a key part of the high tech sector, of course, yet the United States still does not have an official measure of U.S. computer industry productivity. There are many technical problems in forming computer productivity measures, and given current federal budget conditions, these technical problems are not about to be solved soon. Thus, the United States faces stiff international competition without productivity and unit labor cost measures for many (if not most) central high tech products and industries.

The lack of good data in this area has not stopped speculation (or, at best, educated guesses). Recent advances in transportation (for example, containerization), communications (for example, satellite telecommunication systems), and information systems (for example, microcomputers) seem to have enhanced the ability of firms to exploit opportunities presented by large international wage, productivity, and unit labor cost differentials.[4]

These international wage trends, combined with relative occupational employment in several high tech industries, indicate (in part) why many high tech imports will do well in the U.S. market. As noted in table 2–3, the high tech employment of such occupations as blue-collar assemblers is quite large. Almost one-sixth of U.S. semiconductor industry workers are assemblers. Given this fact, many foreign low-wage labor markets, which can provide good semiskilled blue-collar workers, should continue to be attractive sites for high tech investment.

But this growth in foreign high tech investment has a ripple effect on the U.S. economy to the extent that it displaces U.S. domestic high tech investment. For example, a loss of fifteen jobs in the computers and peripheral equipment industries will result in the loss of about twenty-five more jobs in the U.S. economy outside of these specific industries. This is a very large indirect employment effect, and it is in the same general range as the indirect employment effect in the motor vehicles industry (that is, for every seventeen jobs lost in the motor vehicles industry about thirty-four more jobs are lost in the U.S. economy outside of this industry) (Belous, 1984a).

Many have looked toward high tech as an engine of U.S. economic and employment growth, similar to the halcyon days of the U.S. motor vehicle or railroad and steel industries. Beyond short-term foreign exchange rate problems, the indications are that in the United States, high tech industries will continue to face rough, long-term international competition. This international competition could continue to exert strong wage restraints on many blue-collar, and some white-collar, occupations that are highly represented in the high tech work force. It does not appear that high tech related international trading problems are about to go away.

Policy Implications

Policy formulation and evaluation contain a good dose of normative economics. Although others may draw different policy implications from the estimates contained in the earlier sections of this chapter, I view the following six points as the most significant conclusions and policy implications based on this information:

1. Except in a few cases (for example, computer specialists), there is **not** a general scarcity of scientists and engineers. The United States does not seem to need a new, large-scale program to vastly increase the size of its science and engineering work force. Programs may be required, however, to improve the quality of this part of the work force, as it faces difficult international competition.

2. It is important not to confuse cyclical labor market problems with basic structural labor market pathologies. Short-term instability in the demand for scientists and engineers has often caused erratic "cobweb" cycles in these labor markets. Policies designed to reduce this demand instability could have a very positive impact.

Such policies might include reductions in the fluctuation of government spending for defense and research and development. Better labor market information could also reduce this short-term instability. The effort to provide better information would not always require massive public and private investments. For example, the Defense Department already runs a detailed Defense Economic Impact Modeling System (DEIMS). DEIMS is a gold mine in terms of high tech labor market and product market information. Unfortunately, it is a gold mine that few have explored—despite the vast amount of DEIMS information available to the public. Also, despite budget constraints, the federal government should expand its productivity and unit labor cost data system to include information on high tech products and industries. Not to do so would be foolish economizing.

3. The biomedical case presented earlier could be representative of a growing number of scientific labor markets (that is, fewer of the "best and brightest" individuals in the graduate degree programs and more working directly for industry). A simple answer that often comes to mind is to increase the compensation of university scientists. The problem with this is that there appear to be real limits to the levels to which most universities can raise compensation. Within this realistic range, I would not be at all surprised if private industries could, and would, increase their compensation levels to maintain their relative pay advantage over universities. The solution to this labor allocation problem might be more along the lines of reducing the institutional barriers between careers in industry and careers in higher learning. There are already numerous examples of how such mixing and reductions in former institutional barriers are benefiting

both universities and industry (National Academy of Sciences, 1983b, 76–87). These partnerships have included both human and physical capital.

4. U.S. high technology industries could find international competition even harder in the coming years than in the recent past. Given current and expected conditions, U.S. high tech exports might show only modest growth— well under the levels hoped for by many decision makers.

Even if one believes that a new, "grand design" public program for high tech labor markets is in order, the political realities make it an unlikely course of action at present. However, more modest proposals could find wide support from different sides of the political spectrum. For example, the recent report from the President's Commission on Industrial Competitiveness contains several pro- posals designed to improve such things as the quality of U.S. scientific and engineering education; the allocational mechanism of government support for research and development; and the U.S. data and analytical capabilities of industrial competitiveness. Meanwhile, several congressional committees have been considering proposals that contain some similarities to the Presidential Commission's suggestions. The "flavor" of these proposals—on which a com- promise could be reached—is far milder than the industrial policies supported by many House members during the 98th Congress, yet it contains more govern- ment sector spices than are often favored by the Reagan administration.

A major problem in the high tech policy formulation process often centers around "buzz words." Whether we like it or not, numerous policies often have an indirect impact on high technology labor markets, and this indirect impact is often not anticipated. The net result can be an "unconscious" industrial policy. For example, long (and expensive) human capital investments are a major feature in many high tech labor markets. In this era of several new, "grand design" tax proposals, little thought has been given to the impact of tax policy on U.S. human capital investment. However, there are indications that tax policy (including proposed consumption taxes) has a significant effect on human capital investment (Gravelle, 1984). There is a real need to make our uncon- scious actions conscious, as they influence high tech industries and their labor markets.

5. High tech industries contain many jobs that are far from glamorous, that are low-paying, and that involve basic health and safety issues (for example, see Bolle, 1985; Bureau of National Affairs, 1985). Unionization has remained low in many high tech industries. Economic and social conditions could very well make efforts to unionize these industries quite difficult within the ranks of blue-collar workers—not to mention white-collar workers (as an example of difficulties, see Miller, 1984, 35).

6. Given the nature of many high tech industries and firms, it might be useful to think in terms of a generalized "profit rate" time path (and its impact on labor markets). Although some firms may diverge from such a path, it is reasonable to expect that many high tech firms will first face an initial period of

low or zero profit rates (or even losses). This might be followed by a period of growing profits; to use the language of economists, many high tech firms may garner significant economic rents (that is, profit rates that are well above the average for an economy). Firms that enjoy significant economic rents often use part of these funds to raise compensation and other amenities "earned" by various employees. Thus, potential labor–management difficulties can be smoothed over, in part, by using a firm's economic rent.[5] However, as this economic rent diminishes or disappears (and there are not those added funds to smooth over labor–management difficulties), a very different environment, or culture, could develop on the human resources front. I am not arguing that U.S. high tech labor markets are about to become dull, nasty, brutish, and short. However, the estimates presented here do raise the question of whether many U.S. high tech firms will be able to continue to finance the human resources environment of the recent past.

As noted, policymakers and others are often drawn into the high tech field because of its supposed "sex appeal." Many of the estimates and policy issues addressed here are far from sexy, but it may be on this less romantic level that the actual problems and solutions can be ascertained.

Notes

1. For background on this model, see Leontief and Duchin (1984), Duchin and Szyld (1984), Belous (1985), and Leontief (1985).

2. For recent analyses of I/O advances, see Miller and Blair (1985) and Gosh and Sengupta (1984). For a comprehensive analysis of the current status of general equilibrium models, see Weintraub (1985).

3. For details of the methodology used for these projections, see Leontief and Duchin (1985).

4. For more detailed analysis of these forces, see Grunwald and Flamm (1985).

5. For methods of analysis of industry and product life cycles, see Porter (1980, 156–88).

3
The Educational Implications of the High Technology Revolution

Lewis C. Solmon
Midge A. La Porte

The dramatic growth of high technology in our economy was initially interpreted as implying that demands for a more skilled and better educated labor force would dominate labor market needs for generations to come. The initial thinking was that those with advanced training in computer science, engineering, and other areas of science would be in great demand, not only in the high technology industries themselves but also in most of the industrial sector, which presumably would use the products of the high technology corporations. Thus, there have been many efforts to expand education and training in science and technology at all educational levels, from the elementary schools to doctoral programs (Education Commission of the States, 1982; Botkin et al., 1982).

Recent analysis has indicated that although the high technology boom will almost certainly increase the numbers of skilled workers needed in the economy, the fastest growing demand is for jobs requiring low levels of skill—janitors, clerks, cashiers, and so on (Riche et al., 1983).

The focus of this chapter will be the following questions: What types of learned skills will be required, not only in high technology industries but also in industries that utilize high technology? How can we assure that candidates for such positions are of the highest possible quality? In what types of institutions should such employees acquire their skills? To what extent are skills needed for high technology acquirable at the secondary school level, in proprietary training institutes, in public community colleges, at the college or university undergraduate level, or in postbaccalaureate training at the master's or doctoral level? Are there particular efficiencies in ensuring that particular types of institutions provide particular types of skills?

Another basic question involves the role of corporate training in preserving and enhancing the quality of high tech employees. That is, in what skill areas are corporations the most logical providers of education for high technology, and to what extent are the current massive expenditures by corporations for in-house training programs the result of the inadequacies of the education system? Do they reflect specific characteristics and needs of high tech industry? If the educational system were able to become more responsive to the needs of high technology industry, would at least some of the training now going on in the corporate sector be more effectively provided by traditional educational institutions?

It is an important reality that today's workers will have a number of jobs with a variety of different skill requirements during their lifetimes. The typical worker will no longer accept a job at the end of his or her formal education and continue in that job with virtually no change in skill requirements until retirement. Workers will change employers, areas of residence, and even general occupations several times during their careers. It is no longer the case that learning stops once formal schooling stops. It seems apparent, therefore, that one responsibility of formal educational institutions is to prepare workers for whatever recurrent education is necessary in today's world. Educational institutions must provide a good general educational background and a set of attitudes that will enable their graduates to constantly reassess the value of their work and, when the need arises, to acquire new and different skills so that their productivity will be maximized.

Before looking specifically at the relationships among the educational system, individuals, and the demands of high technology industries, we need to focus briefly on the role of research and development in influencing demands and competencies.

Research and Development

It is useful to begin our analysis by looking at some statistics that briefly summarize high technology problems in the United States (National Science Board, 1983). In 1965, the United States had approximately 494,500 scientists and engineers engaged in research and development (R&D), as compared with 521,800 scientists and engineers in the Soviet Union. Because of the Soviet Union's much larger population, the proportions of scientists and engineers engaged in R&D in the labor force differed as follows: 64.1 per 10,000 workers in the U.S. and between 45 and 48 per 10,000 in the Soviet Union. In addition, the United States had more than four times as many scientists and engineers engaged in R&D as Japan did; and about ten times as many as West Germany, the United Kingdom, and France.

By 1982, the number of scientists and engineers engaged in R&D in the

United States had grown to 716,900 (a 45 percent increase), whereas the number of scientists and engineers in the Soviet Union had grown to 1,340,400 (a growth of 157 percent). Over the past fifteen years, the growth rate of U.S. scientists and engineers engaged in research and development has been substantially slower than that in the Soviet Union, the United Kingdom, Japan, West Germany, and France. Also, the share of gross national product devoted to civilian research and development expenditures has grown more slowly in the United States than it has in France, West Germany, Japan, and the United Kingdom (National Science Board, 1983).

These statistics leave little doubt about why U.S. superiority in high technology has diminished compared to the other countries. Once the post-Sputnik growth of interest in science and technology waned, the enthusiasm for and commitment to research and development slowed dramatically.

Within the United States, there has been a dramatic change in the patterns of funding and the conduct of research and development. Between 1960 and 1983, total real national expenditures for research and development grew by 101.7 percent. However, the growth in expenditures for research and development by industry (as opposed to government and academe) grew by 208.9 percent. Similarly, while total expenditures for industrial R&D grew by 92.6 percent, such expenditures from nonfederal sources grew by 210 percent. Despite this growth in the share of research and development funded by the industrial sector, the share of R&D actually performed by industry fell from 77.9 percent to 74.4 percent over the past twenty-three years (National Science Board, 1983).

Research and development expenditures are generally divided into three categories: basic research, applied research, and development. Basic research expenditures grew 175.9 percent between 1960 and 1983; applied research expenditures grew 106.1 percent; and development expenditures grew 90.8 percent. This resulted in approximately a 50 percent increase in the share of basic research (from 8.8 percent of the total to 12 percent of the total). Applied research has stayed relatively constant at about 22.5 percent, and development expenditures have fallen from 68.9 percent to 65.2 percent. Industry's funding of all research grew by 151.5 percent between 1960 and 1983, compared to an overall growth rate of 125.9 percent.

Funding for research by industry has been focused more on applied research than on basic research. Industry has also undertaken a much greater share of expenditures for development. American industry has begun to shoulder an increasing portion of the funding for research and development at all levels, with the possible exception of basic research. As the nation begins to rely more and more on corporate-sponsored research, the often-expressed fear that basic research will suffer compared to applied research and development takes on more credence. It is also true that as a larger share of research, particularly in the applied and developmental areas, is undertaken in industrial settings, the need for highly skilled labor expands in that sector. Certain industries—such as chemicals and allied products; nonelectrical machinery, particularly office computing and accounting machines; electrical equipment, particularly communications equipment; motor vehicles and other transportation equipment; aircraft and missiles; and professional and scientific equipment—account for the bulk of the R&D expenditures by industry.

Thus, there has been rapid growth in national expenditures for R&D, of which an increasing share has been funded by the private industrial sector. We also recognize that growth has been focused more on applied research and development than on basic research. Yet despite this growth, our relative advantage vis-à- vis other nations has been declining.

The main focus of this chapter is on the situation in the academic and corporate sectors regarding the provision of the highly skilled manpower necessary for growth in research and development. Of particular concern is the erosion of math and science education in the secondary schools.

Secondary Schools

We will begin by looking at the precollegiate sector of U.S. education, particularly the secondary school. In 1976, 31.3 percent of public secondary school teachers indicated that they spent the largest portion of their time teaching mathematics or science; by 1981, 27.5 percent of teachers so indicated. Between those years, total enrollment in grades 9 through 12 fell from 15.7 million to 14.4 million (U.S. Department of Education, 1983). By analyzing these figures, we find that the absolute number of high school students being exposed to math or science specialists fell by 20 percent between 1976 and 1981.

As enrollments decline, for our high schools to send the same numbers of students into postsecondary education with math and science backgrounds as solid as they had been in the recent past, we will need a greater concentration of math and science teachers in our high schools. Yet as labor demands and salaries of scientists in industry rise, secondary school teaching becomes less attractive for scientifically oriented college graduates. By satisfying the short-term needs of industry, we may be hurting our technological competitiveness in future generations.

High school seniors in 1980 had taken the following courses: algebra II, 49 percent; geometry, 56.2 percent; trigonometry, 25.6 percent; calculus, 7.8 percent; physics, 19.4 percent; chemistry, 37.3 percent (National Science Board, 1983). In other words, few of the nation's high school students are getting a reasonably advanced math/science education. Even if all of them go to college, less than half of those entering college do so with a solid math/science background.

In *Science Indicators 1982*, the National Science Board (1983) illustrates the declining state of precollegiate mathematics and science education:

There was deteriorating performance of all precollege students on math and science achievement tests.

There were declining scores on tests used for college admissions, such as the SAT.

A 15 percent drop in the number of 18- to 24-year-olds was anticipated during the 1980s.

There was a fear that science and engineering might not be attracting as many highly able students as other professions are.

Between 1972 and 1980, the proportion of high school students in academic (versus general or vocational) programs declined by four percentage points; in 1980, only 39 percent of high school seniors were in academic programs. This may be more of a problem when the number of high school students declines during the latter part of the 1980s. Students in academic programs took more course work in math/science than students in other programs did. For example, more than half of the seniors in academic programs in 1980 had taken three or more years of math, compared with 18 percent of those in vocational programs.

The National Assessment of Educational Progress (NAEP) showed average math skill levels of 13- and 17-year-olds declining during the 1970s, with 17-year-olds showing the largest drop. Similar negative results were reported regarding math understanding and math applications, as well as science achievement.

It is clear that during a time when high technology and its concomitant labor demands are growing, we are facing a situation in our secondary schools in which students are becoming less well trained in the very subjects that will feed the high technology industries. This is a problem not for academe alone. The industrial sector must begin or continue thinking about ways to stimulate mathematics and science education of the highest quality possible at the precollegiate level. This might involve providing funds to attract and maintain quality teachers, to upgrade laboratory and other teaching equipment, and in general to help provide facilities and faculty that will both attract and stimulate the best and the brightest of our students.

More creative interactions must also be considered. One approach might be development of ways to provide corporate employees as teaching personnel for the secondary schools on a revolving basis. It is very difficult for corporations to constrain their hiring of new college graduates so that sufficient supplies of science and engineering majors will be available to teach in the elementary and secondary schools. The manpower is needed in industry, and industry should be expected to do the best it can to attract the highest quality talent. However, this might be a short-term solution if, by depriving the high schools of adequate teachers of science and mathematics, these same corporations find themselves, several years hence, without sufficient numbers of scientists or if candidates are below the level of quality the industries require.

The problem with expecting corporations to restrict their hiring of bachelor's, master's, or doctoral degree recipients in science and engineering so that various levels of educational institutions can be assured of having adequate faculty is an example of what economists refer to as the "free-rider problem." In essence, all corporations might agree, in spirit, to regulate their hiring to achieve this goal. However, once the agreement is reached, each corporation has an incentive to violate it. By having all other firms restrict their hiring, the firm that violates the agreement will be able to staff its own needs adequately while being assured that there will still be sufficient faculty to produce the next generation of scientists and engineers. It is then possible that each individual firm would make the same decision; therefore, all firms would violate the agreement, which would thus fail. It is very difficult to get profit-seeking corporations to agree on a policy that is against their individual short-run self-interest.

Prebaccalaureate Institutions

The postsecondary sector of our educational system below the baccalaureate level is an important source of personnel for high technology industries. In particular, data seem to indicate that two-year colleges are becoming more technical in orientation. In 1970–71, of 307,880 degrees awarded by two-year colleges, 94,621 (31 percent) were in technical fields (data processing, health services, mechanical/engineering technologies, and natural science technologies). By 1981–82, the number of such degrees had risen to 236,378, or 42 percent of the total (U.S. Department of Education, various years).

The two-year college sector is among the most rapidly growing in U.S. education. It is often viewed as a second-class citizen, however, in the sense that most of the brightest high school graduates proceed directly to four-year colleges. Students in the two-year colleges are predominantly from lower socioeconomic status families who are unable to afford the tuition, forgone earnings, and mobility that are often required to enroll in a four-year institution. Two-year college graduates provide the manpower for middle-level technology jobs, as opposed to jobs in science and engineering, which will become increasingly important in high technology industries (Silvestri, Lukasiewicz, and Einstein, 1983). Again, this leads to the inference that the high tech corporate community should take an increasing interest in the community college sector, not only by providing financial and political support but also by working closely with the two-year colleges to make sure that they will provide the kinds of manpower specifically needed by the high technology sector.

A large but often ignored sector of U.S. postsecondary education is the noncollegiate, noncorrespondence postsecondary schools with occupational programs. These are often referred to as private, for-profit, or proprietary institutions. Because they generally are profit-seeking, they are much more likely to be responsive to the needs of industry than are other types of institutions that are embedded in complicated, politically constrained bureaucracies. Many of the so-called proprietary schools are small, are run by a few individuals, lack tenured faculty, and are able to adapt and readapt their programs on relatively short notice. In addition, many of these schools do not restrict their programs to fixed time periods, such as two or more years; rather, they can turn out technicians and others with very specific skills in a short time period. The proprietary schools do not generally require that their students take a variety of general education courses; they allow them to focus on skill training. Their flexibility in scheduling courses also more readily accommodates students who work full-time.

In 1975, of the 1,399,100 students enrolled in noncollegiate, noncorrespondence postsecondary schools with occupational programs, 53.3 percent were in vocational/technical (voc/tech) schools, technical institutes, or trade schools. Despite a growth in total enrollments to 1,687,000 by 1980–81, the total percentage fell to 45; enrollment in voc/tech programs fell by 5.1 percent and enrollment in technical institutes fell by 33.6 percent. Trade school enrollments saw a growth of 44.4 percent (U.S. Department of Education, 1983). If we assume a graduate-to-enrollment ratio of approximately 65 percent for proprietary school programs that last six to twelve months (Herbert and Coyne, 1980), then approximately 493,450 new technology graduates were produced in these institutions in 1980–81.

It is unclear why enrollment in institutions that focus on vocational and technical training declined over the past decade. It is possible that a declining interest resulted from the increase in in-house education by private corporations. It is also possible that occupational training for high technology industries is so costly that it is not cost-effective for many of the proprietary schools. In any case, serious study is necessary to determine whether there is untapped potential in the private, for-profit postsecondary institutions as a training base for the technicians who will be increasingly demanded by the high technology sector of private industry.

Perhaps the most important message deriving from the foregoing discussion is the need for high technology industries to maintain and increase their interests in and interactions with the prebaccalaureate institutions of postsecondary education. Some high tech corporations—for example, Bell and Howell—have gone so far as to develop their own proprietary schools or purchase existing ones. By focusing almost exclusively on the baccalaureate-granting colleges in this country, high tech might be missing an opportunity to have an influence on those institutions that provide a significant and increasingly important portion of their labor needs.

Baccalaureate and Postbaccalaureate Institutions

Although the technicians prepared by the aforementioned prebaccalaureate postsecondary institutions will do a great deal of the work in industries that use high technology, the competitive edge of high tech companies, both domestically and internationally, will in all likelihood be achieved by the productivity of those with baccalaureate or higher degrees. Therefore, we will discuss the availability of scientists and engineers who hold at least a bachelor's degree and then the implications for our nation's colleges and universities regarding the changing demands for such graduates.

The most recent data are generally available only for 1980–81; therefore, we will compare degrees awarded between 1972–74 and 1980–81. The total number of bachelor's degrees awarded declined between 1973–74 and 1980–81 from 945,776 to 935,140 (U.S. Department of Education, 1976–83). In 1973–74, 15.5 percent (146,195) of the degrees awarded were in the biological sciences, computer information sciences, engineering, math, and physical sciences. By 1980–81, 18 percent (168,367) of all degrees were in those fields. Patterns within fields have not been consistent. The number of degrees in biology declined from 48,340 to 43,216; in math, there was a decline from 21,635 to 11,078. Mathematics has seen a steady decline from 1973–74 to the present. The greatest growth was in the computer or information science area—from 4,756 to 15,121. Physical sciences grew slightly, from 21,178 to 23,952. The peak year in degrees awarded in the biological sciences was 1975–76, when 54,952 degrees were awarded.

The total number of master's degrees awarded between 1973–74 and 1980–81 increased from 277,033 to 295,739. However, the share of those degrees in the aforementioned science fields fell from 12.7 percent (35,103) to 11.8 percent (34,756). There were fewer master's degrees awarded in 1980–81 than in 1973–74 in the biological sciences, mathematics, and the physical sciences, whereas the number of degrees awarded in computer or information science and engineering rose.

At the doctoral level, the numbers of degrees awarded declined slightly from 33,816 in 1973–74 to 32,958 in 1980–81. More recent figures show that in 1983, 31,190 doctoral degrees were awarded (National Research Council, 1983). The number of doctorates in the science and engineering fields previously noted fell steadily, from 11,606 in 1973 to 10,400 in 1980–81 and 9,655 in 1983. Doctorates in engineering, mathematics, and the physical sciences have declined each year since 1973–74.

It is very striking that the proportion of recent graduates who are employed in business and industry—as opposed to other sectors of the economy, such as government or, in particular, academe—has been growing in recent years. For this analysis, we will refer to 1974 and 1975 graduates and their employment situation in 1976 as the "early cohort" and 1978 and 1979 graduates and their employment situation in 1980 as the "recent cohort." At the bachelor's level, 48.9 percent of the early cohort science graduates were holding jobs in business and industry, in contrast to 57.5 percent of the recent cohort. The figures for engineering were 79.4 percent of the early cohort and 87.7 percent of the recent cohort (National Science Foundation, 1982, 1983, and unpublished data). It is clear that most bachelor's level engineers have traditionally entered business and industry. However, the increase in scientists in the business sector does not portend well for the supply of science teachers, particularly at the secondary level. Moreover, as more and more baccalaureate recipients move directly into industry, the pool for master's and doctoral students decreases.

At the master's level, 23.8 percent of the early cohort of scientists took jobs in business and industry, but this rose to 39.8 percent for the more recent cohort. Among master's level engineers, 64.4 percent of the early cohort and 75.8 percent of the recent cohort took jobs in industry. At the doctoral level, 14.7 percent of the scientists and 56.1 percent of the engineers in the early cohort took jobs in industry; these figures rose to 21.0 percent and 64.3 percent, respectively, for the recent cohort. The move to industry by doctorate recipients might be explained by the declining academic demand for faculty, even in science, because of the decreasing size of the age cohort. However, more attractive job offers from industry are an additional explanation for the flow of doctorate recipients to that sector.

By the mid-1990s, academic demand will once again increase as the children of the "baby boomers" reach college age; demands in high technology industry will also continue to grow. In the next few years, so long as the demand for faculty remains weak, high tech industries will be able to satisfy their demands for skilled scientists by taking a growing proportion of a constant or even declining pool of recent degree recipients. However, once the demand for faculty begins to grow again, the competition between academe and industry for highly trained scientists and engineers could become disruptive. As noted earlier, the current efforts by industry to attract more highly skilled baccalaureate-level scientists and engineers does great disservice to those who are trying to restock our nation's secondary schools with capable science and mathematics faculty to prepare the next generation of college and graduate school students.

Another concern that both the academic and the business communities must keep in mind is that the success of our high technology industries will depend not only on the quantity of scientists and engineers but also on their quality. During the 1970s and early 1980s, the demand for graduates with MBAs, law degrees, and engineering degrees was very strong. Given these demands and a variety of other factors, we have seen dramatic shifts in choice of major by new college students. An earlier study shows that, in particular, there were enrollment share declines in the arts and humanities, education, social sciences, professional fields, and—particularly relevant for this discussion—the physical sciences (Solmon and La Porte, 1986). Moreover, the share of high-quality students grew most dramatically in the fields of business and engineering but fell precipitously in the arts, humanities, education, the social sciences, and the physical sciences. We counted the total pool of highest-quality students at 70,274 in 1967 and 154,892 in 1983. There were approximately 4,000 fewer high-quality freshmen intending to enroll in the physical sciences in 1983 than there were in 1967. This compares to an increase in the numbers of high-quality students in engineering of 22,700 and in business of 19,500.

In attempting to rank the strength of eight fields in terms of the quality of students enrolling and the changes in those figures between 1967 and 1983, we found that the strongest field was business, followed by engineering, the biological sciences, professional fields, education, and the social sciences, and only then by the physical sciences, the arts, and humanities. The strength of the physical sciences, as indicated by the quantity and quality changes in freshmen indicating an interest in the fields, is among the lowest of any fields offered by most colleges and universities today. Thus, despite the fact that, as we have already indicated, bachelor's degrees awarded in the physical sciences rose from 21,000 to 24,000 between 1973–74 and 1980–81, we must recognize that the physical sciences seem to be weakening in both quantity and quality compared to other fields.

It is probably going to be necessary to provide financial incentives in the form of fellowships and scholarships to bright high school graduates to attract them into the physical sciences. Perhaps contributions to the fellowship pool might be obtained from private corporations, particularly if some quid pro quo were available to them. This might be in the form of preferential access to graduates for companies that contribute. Although the engineering fields look strong again, it seems that we might be sacrificing long-term basic research and development in favor of more pragmatic applications. This is a short-run solution to a problem of long-run competitiveness for the high technology fields. Without a continuing adequate supply of scientists, our high technology industries could be in bad shape for decades to come.

Corporate Education and Training

In assessing the educational implications of the high technology revolution, it becomes clear that an individual's education can no longer stop after he or she has ceased to be enrolled in an educational institution. In many fields, a professional's education now has an estimated half-life of only five years; some estimates put that figure even lower. Thus, the demand for continuing education will continue to increase. The boundaries of the term *educational institution* have, by necessity, been broadened, and the employment setting has assumed some of the functions of traditional institutions. We now turn to a discussion of the continuing education and training that goes on in the corporate sector, with particular reference to high technology companies and the new strategies employed by them to increase employee productivity.

Some have questioned whether, if profit margins are reduced in high tech firms, companies will continue to be able to finance educational opportunities for their employees. If that were the case, one would expect to see a reduction in the general education or "paternalistic" human resources development programs some corporations have established—those for which the links to quality and productivity are not so obvious. Yet most firms will not cut back significantly on their continuing education programs in vital areas such as technical fields and management. What we may see, instead, is an increased emphasis on maximizing the investments in training. That includes developing better ways of quantifying the outcomes of training programs. There is a well-developed theoretical base for calculating rates of return to other forms of education, and this needs to be extended to employee education. It may be to higher education's advantage to develop cost-effective methods of educational delivery for business and industry. If educational institutions can provide low-cost, efficient means of delivering educational services to companies that choose not to maintain full-time training staffs or decide not to operate certain general education programs, then companies may turn to them for these courses.

Training and Productivity in High Tech Firms

With an increased emphasis on productivity improvement and quality—of both products and processes—there is an increased need for continuing education for employees. There is some evidence of a link between training and education and improved productivity. In a 1984 study of productivity improvement efforts by U.S. industries, conducted by the Bureau of National Affairs, the results of the use of training and education to increase productivity were impressive, with 77 percent of the 195 firms in the study reporting that training programs were "highly effective" or "overall encouraging." The results of such training programs can be diverse, including such things as improved communication or improved quality of work. More companies are instituting new training and development programs for employees than any other form of organized productivity effort (Bureau of National Affairs, 1984a). In addition to productivity improvements, corporate education and training programs are significant in preserving and enhancing the quality of employee skills—from the technician to the Ph.D.

In general, *continuing education* means educational activities engaged in after full-time professional employment has begun and includes courses that update one's knowledge in a current specialty or develop expertise in a new field. A 1979 study by the New York University Center for Science and Technology Policy found a positive relationship between company expenditures for R&D

and continuing education activities and between training and education and staff R&D performance. Hewlett-Packard, for example, invests 10 percent of its net sales in R&D and supports that $400 million investment with a well-developed corporate educational system.

In high tech industries, recurrent education and training opportunities are considered a key factor in maintaining and stimulating vitality and innovation among scientists, engineers, and other technical personnel. Without enhancing the knowledge and stimulating the creativity of scientific personnel, the value of the investment companies make in hiring top-quality personnel could be seriously diminished. Real technology-driven skill shortages exist in the quality, not the quantity, of workers. Eventually, employers will be able to draw from a pool of engineers, but constant, job-specific retraining will be required. Employee job training requirements are closely connected to the nation's economic adaptability (Gorovitz, 1983).

Education and training within large private sector corporations in the United States has increased dramatically over the last decade. Millions of adults, as employees, are involved in formal and informal training programs. America's workers and managers have been enrolled in continuing education programs in the past, but in recent years their numbers have increased, the variety of subjects has broadened, and, most interesting, corporations have become their own well-organized educational providers (Levine, 1982; Eurich, 1985).

Corporations are spending upwards of $30 billion to $50 billion annually to educate and train their employees. Precise amounts are difficult to determine, either because the information is considered proprietary by individual companies or such data are not kept. Thus, although reliable figures are hard to ascertain, the basic fact is that U.S. employers spend on in-house programs a sum comparable to the combined appropriations for higher education in all fifty states, which in 1980 added up to $20 billion. Employee continuing education has been broadened to include such nontraditional modes as teletronics, correspondence, computer-based instruction, and corporate-based education.

Although education in industry is increasing across the board, high tech firms in particular appear to engage in extensive training programs for employees. Some have suggested it is because of their "excess profits." Yet we believe there are certain characteristics of high tech industries that necessitate employee training and education. Rapidly changing technology, short product life cycles, and rapid skill obsolescence are key factors. Also, competition for good technical personnel increases the pressure on firms to provide a wide range of incentives to encourage innovative personnel to stay.

Educational benefits, including continuing education and paid sabbaticals, are used particularly in geographic regions where competition for good technical personnel is extreme. Some companies make these benefits broadly available; others provide them as a privilege only to those selected. In progressive high tech firms educational benefits have replaced, in some respects, traditional "perks," such as plush offices and executive parking places. Such educational benefits are seen as more important to the success of the firm, as well as to the advancement of the individual, than other fringe benefits.

The popular perception of high tech scientists and engineers is that they are risk-loving entrepreneurs who are eager to start up their own firms. However, there is some evidence (Smith, 1985) which suggests that, for the most part—given the proper incentives (which also include financial incentives, such as targeted profit sharing and bonuses)—most skilled personnel prefer employment security to speculative venture. Yet a problem most firms face is the pirating of their technical personnel by other companies. Some companies have adopted a cautious approach, requiring such things as employment compensation agreements or "contracts" to serve in the company for a certain number of years as a precondition for participation in employer-sponsored education. A few have even cut back on the depth of their training opportunities for fear of losing highly trained personnel and, possibly, company information.

This appears to be an unwise strategy. Trying to limit the depth or extent of knowledge of scientific and engineering personnel not only hurts the firm in terms of productivity and innovation but may also actually encourage workers to leave. Other companies, which provide ongoing skill advancement opportunities, may be seen as more progressive and more interested in employee development. Educational opportunities are viewed as a fringe benefit, and a lack of such opportunities may hurt the company's recruitment efforts. A 1981 study by Mountain Bell/AT&T found that employees who completed their degrees while employed at a company were more than three times as likely as other employees to remain with the company for a significant period of time (Huddleston and Fenwick, 1983).

The problem of labor mobility is a real one, nonetheless. Highly educated workers are generally more mobile than less-educated workers. Whether or not company-provided training increases worker mobility or enhances retention will have to be studied further. In the meantime, it appears that most high tech firms are proceeding with the development and implementation of training programs as part of a long-term strategy to retain skilled personnel.

Innovation is essential in high tech companies. In an organization that relies on the creativity of its employees, particularly at the higher skill levels, there is a need to regard humans as capital—to develop a human resources accounting system that operates in somewhat the same manner that cost accounting works regarding physical resources. One company, Dana Corporation ("Dana University," 1977), has developed a computerized system for cross-referencing all employees by the type of training they have received. Thus, when a position becomes available in the firm, managers can readily see which employees have the requisite training, enabling them to promote from within the firm.

Many of the training programs in high tech industries are significant in that they represent a growing trend toward carefully constructed and integrated systems of education for employees. They are not merely ad hoc training programs—they emphasize the need for continuous upgrading of employee skills; they provide a vehicle for employees to participate in the design processes of work; and they offer forums for employee creativity.

By diversifying the skills of the work force so that each worker can perform a variety of tasks, greater flexibility is created. Thus, when technology changes and production processes shift, workers can more readily be shifted where they are needed and can be better prepared attitudinally for eventual shifts.

Also, if workers are more knowledgeable about the overall production process and management goals, they may be better able (or management may perceive them to be better able) to contribute effectively to decision making. Training and education can be viewed as a precondition for effective participation by the work force. The process of consultation with employees can be used as a technique for solving technical labor planning and products problems. The expertise of employees can thus be used by the company as a potential cost-cutting and labor-saving device. It may also increase worker job satisfaction and morale. Training thus should be conceived of and implemented as a step in an overall process of involving employees in the production processes of the firm.

There is some evidence (Tsang, 1984) that training's potential for increasing worker productivity and job satisfaction is great, but if the training is not accompanied by a plan to involve workers in decision making (at least through a consultative process), the firm can be in worse condition. The problem of underutilization of education in the United States may have been a significant factor in what has appeared to be a long-term decline in productivity in the economy. One can draw attention to the Japanese experience, in which a rapidly expanding educational system is matched with a production system that involves workers in production decisions (Ouchi, 1981). Part of the success of Japanese companies may be a result of their ability to utilize educated labor more fully (Tsang, 1984, 15).

The problem is not "too much" education but, rather, the underutilization of worker skills and knowledge in production decision making. A critical incentive and reward for an experienced engineer or scientist is the knowledge that he or she is influential in the decision-making process. If managers make all the decisions solely on the basis of the authority of their positions, this has a stifling effect on the entire organization.

In high tech firms, although most training may be focused on technical personnel, management training is not and should not be neglected. A 1982 study revealed that in all industries, 40 percent of all courses were concentrated in business and management (including subjects such as managing time, managing people, fiscal management, and managing production and operations).

With so much time, in-house personnel, and corporate dollars being devoted to managing people through such courses as "Team Building," "Effective Listening," and "Motivating Others," a question for traditional educators may be posed: "If corporate education must spend so much time on personal relationships and team action, have schools and colleges been so individually-oriented that students have not learned to work with others?" (Eurich, 1985, 65)

Although many management courses are taught in-house, outside experts, rather than in-house staff, are usually utilized at top management levels. Seminars may be conducted at a university campus or special institute. Executive programs frequently are of longer duration than other company training programs. Some programs are less company-specific and the context is broader; such programs can be important to the vitality of the firm. In recent years, executive education programs have been developed at many universities around the United States. Systematic study of the content and relative quality of such programs should be undertaken.

Public affairs is often a focus of executive management courses in high tech firms because of the challenges posed by international competition. In a Conference Board survey of 176 firms among *Fortune*'s top 1,300 industrials and nonindustrials, it was found that nearly half of the corporations provided courses in public affairs (Lusterman, 1977).

One problem is that many corporate education programs tend to separate training for management from training for technical personnel. Some division may be justified, but the separation can also narrow the company's vision and create problems for the future leadership of the firm. James P. Baughman, director of management education for the General Electric Company, stated: "We find ourselves in a kind of no-man's land in the curriculum between the engineering and science people who really have the technical ability but can't place it in a management context and the people who have management sensitivity but really can't link it up with technical know-how" (Business–Higher Education Forum, 1984). Separating management education from technical education produces corporate leaders who lack familiarity with the whole organization. It can also produce executives who are not in touch with day-to-day problems of the technical work force. This can be a particular problem in high tech firms, where long-range planning is influenced by rapidly changing technology.

Effective educational systems also include training workshops conducted by one group of employees for other groups. For example, at Tandem Computers, Inc., technologists regularly train administrators on the performance and function of the firm's products and, in turn, administrators train the technologists on personnel policies and financial operations (Maidique and Hayes 1984, 56). This interaction is significant, because employee training programs for managers and technical personnel are sometimes unnecessarily kept separate. Thus, one group often knows little about the other's problems and concerns. This lack of knowledge about the whole process may hamper intergroup relations and, ultimately, company productivity.

As the emphasis on quality and productivity of employees increases, the need to enhance worker capabilities on all levels will grow. However, there may be a greater need for evaluation of training programs—much more than has been done in the past—and the need will be for formal evaluation, not the random, word-of-mouth evaluation that usually constitutes the only evaluation in which a firm engages.

Because their cost-effectiveness is difficult to measure, educational programs in industry are sometimes considered a luxury. They are often the programs cut first when profits falter and budget reductions become necessary. Educational programs are often funded through overhead. Moreover, some industries still believe that education is the sole responsibility of the employee. But a growing number of companies—high tech firms are among the leaders—realize the benefits of employee education and training to them as well as to their work force.

The number of firms engaged in formal training is increasing rapidly. The depth and breadth of course offerings and the extent of employee involvement are also growing significantly. Observers of this phenomenon, believe that it is not an altruistic gesture or a passing fad but, rather, a bellwether for the future. Continuing education, training, and retraining are necessary preconditions for employee and corporation productivity, and it is for this reason that corporate involvement in training and education is expected to continue and to grow. Although it is quite likely that the cost/benefit ratio of technical education will make it a continuing necessity in the corporate sector, programs involving general education or personal development may be more reasonably viewed as fringe benefits to be cut when a profit crunch occurs. And it is widely acknowledged that basic skills training should not be a function of corporate education. This is clearly a job for secondary schools or even elementary schools. Thus, if corporate America does not wish to be faced with providing such training, it is in its interest to see that public education does its job.

So long as shareholders and profits are given primacy in the minds of corporate decision makers, corporate education will not become a substitute or replacement for traditional public education. Too many objectives of the latter are only remotely related to profits and so will not be taken over by corporate education. The most we can expect is that the corporate sector will focus on technical updating and training relevant to its production needs and on general education to the extent that it reinforces the productive capability of its workers. Industry will also continue to provide traditional sales and customer training as well as management training. Anything beyond that will be viewed as a fringe benefit and will be treated as such.

Conclusions

This chapter has suggested a number of ways in which educational institutions can meet the needs of high technology industry. Basically, interaction and cooperation between industry and academe must be greatly expanded and encouraged. Educational institutions at all levels must make significant efforts to determine what the short- and long-term needs of industry are in terms of scientific and engineering manpower. Similarly, those in the high technology sector must communicate with those in the academic sector to inform them of the prospective labor demands for the foreseeable future. Although the usefulness of many of the labor market forecasts of the past has been justifiably questioned, it would be useful to launch a cooperative effort to develop an information system that would enable both the academic and the business communities to track future needs for scientists, engineers, and technicians.

Such a system could address the question of the extent to which a college education will be necessary for upcoming age cohorts. Over the past several decades, a large percentage of families in the United States have come to assume that a college education is the minimal educational achievement a youth should seek before entering the labor market. It has reached the point that everyone thinks that he or she should become a medical doctor, a lawyer, or an MBA. Of course, experience has shown that for many people, such aspirations are unrealistic. Perhaps the real function of the educational system is to get students in the system, particularly at the secondary level, to recognize, once again, the dignity of traditional labor. Secondary schools must alter the expectations of the population, which increasingly eschews traditional work for anticipated careers in "the professions." We may have a problem in the future with the availability of low-skilled workers, particularly with regard to the satisfaction of those who end up in such jobs, unless the expectations of the vast majority of our youth are altered.

One approach is to provide for different treatments for different people—that is, to identify those who would serve the economy most productively as professionals and those who would serve the economy most productively in more traditional skilled and unskilled labor. This approach is subject to the charge of elitism, or worse. However, we must recognize that more education is not always the best course for everyone. There are many productive ways to serve the economy that do not require high level academic preparation.

Current demographics make these types of arguments more compelling. As the secondary school-age cohort gets smaller, shortages are bound to arise in the types of jobs traditionally held by those who obtain secondary education and no more. Thus, salaries of such students will rise, and the opportunity costs for people going to college will also increase correspondingly. With higher opportunity costs, the number of those entering the labor force directly after high school will increase.

Vocationalization of a curriculum depends on the extent to which such a curriculum is viewed as terminal in the educational process; in other words, the curriculum that serves as the entry point to the labor market generally becomes the most vocationalized. We have observed this in recent years; as the baccalaureate degree has become the standard terminal degree, students have been more likely to select programs in business or other vocationally oriented fields. If more students tend to enter the labor force directly from high school, it is necessary that the high school curriculum orient itself more toward preparation for work and less toward preparation for academic study beyond high school.

At all levels of education, from secondary schools on up, there are current or prospective problems of staffing courses in science, mathematics, and engineering. Industry must be more than a competitor with academic institutions for scientists, engineers, and technologists. Industry must develop a greater sense of responsibility for the staffing of academic institutions, rather than trying to hire all of the best people away from academe. These hiring attempts, particularly in the high tech fields, are usually successful because high tech can offer higher salaries than academic institutions.

A sufficient supply of teaching personnel could be guaranteed, in part, through a "lend-lease" program, whereby corporate employees would spend a certain number of years in the classroom teaching the next generation. Perhaps a variety of exchange programs would serve not only to provide manpower in the classrooms of the nation but also to enable teachers to gain insights into the needs of high technology industry by working for periods of time in the corporate setting. Another advantage of these types of interchanges might be to average salaries, over a number of years, so that teacher salaries and salaries in the corporate sector would not be so disparate as they currently are. Teachers could most readily do this during the summer months, when no classroom time would be lost. Provisions would be required in any such exchange programs to ensure that teachers would actually return to their classrooms after spending time in a corporation. This suggestion applies equally well to the secondary school as to the university. Just as there are shortages of science and mathematics teachers in the secondary schools, there are shortages of professors of engineering and computer science in most universities around the country. Appropriately designed exchange programs could go a great deal of the way toward rectifying such shortages.

As the availability of teachers and professors becomes more critical, particular efforts will have to be made at all levels of the educational system to augment academic salaries so that the best teachers can be retained within the educational system. This, too, could involve education–industry cooperation. It is possible that rather than relying on public tax initiatives or federal subsidy, another more promising method of assisting education could be to develop some sort of industrial contribution to the merit salary pool of educational institutions, perhaps in the form of endowed chairs or similar mechanisms.

Another area for cooperation between the educational institutions and high technology industry is in the funding of research. It is clear that the preponderance of research conducted in the industrial sector is oriented more toward applied research and development than toward basic research. Yet it can sometimes be forgotten that unless basic research is expanded regularly, a time will come when applications derived from basic research, and development of commercially viable products from such applied research, will be stunted. Thus, as in many other aspects of U.S. industrial life, corporate leaders must take a long-term view; that is, at least some part of corporate expenditures on research and development should be devoted to basic research that might more appropriately be conducted in university laboratories. Although this might have only indirect payoffs to high tech firms in the short run, it is absolutely essential over the long term. If such a reallocation cannot be achieved voluntarily, perhaps a tax on research spending by corporations, which would be devoted to basic research, would get around this "free-rider" problem.

There are certain initiatives that are clearly the sole responsibility of educational institutions vis-à-vis the science, engineering, and technical labor force and the high tech industrial sector in particular. First, educational institutions at all levels must recognize that scientists, engineers, and technicians will necessarily have to continue learning well beyond their graduation from formal schooling. Thus, whatever can be done by the schools and colleges to prepare their graduates to continue learning after graduation should be encouraged. Training must be flexible so that it can be augmented throughout the life span of the individual.

In addition, the orientation of even the most focused technical training must be broadened to include such skills as critical thinking, problem solving, writing, and communication skills, which are ultimately necessary in a wide variety of science and engineering jobs. Our research has indicated that the normal career path for the majority of scientists and engineers is to move into management positions after a certain period of time. If this continues to be the case, those who train scientists and engineers should recognize that some basic understanding of business principles, accounting concepts, and personnel relations might be valuable add-ons to technical graduates' skills.

Educational institutions, particularly those at the postsecondary level, must be cognizant of the fact that students who are capable of studying in high technology fields are incurring substantial opportunity costs by continuing their studies rather than going directly into the labor market. Thus, these institutions must take every possible step to minimize the opportunity cost of obtaining such training. This would involve consideration of when courses are offered (perhaps at night, on weekends, or during vacations) and how long they continue. Some reconsideration of the traditional time elements in education is necessary. It may be that certain kinds of skills and certain kinds of degrees can and should be earned not in the classical two- or four-year period but in a shorter period of time. Similarly, extensive efforts—perhaps again by collaboration between industry and education—must be made to develop fellowship and scholarship opportunities, particularly for those training in the sciences and engineering and technical fields so that their costs of remaining in school are alleviated. The social value of an adequate supply of highly able scientists and engineers is certainly great enough to justify substantial subsidy of those who are training for these fields.

Such subsidies are especially important in attracting the highly able students who tend to have the best alternative opportunities.

If we look at a snapshot of the labor market for scientists, engineers, and technicians in high technology industries today, we see a fairly satisfying picture. Computer science and certain subfields of engineering are in short supply; however, in most fields, high technology firms can be confident that they will be able to hire the numbers of scientists, engineers, and technicians that they need. The question of quality, however, could soon be upon us, and our ability to forecast the state of demand/supply equilibrium beyond this decade is precarious at best. The one thing that does seem clear is that we will best be able to meet the challenges of providing and keeping a technologically adequate labor force if those in education and those in the high technology industries interact more often and at higher levels, so that planning can be done cooperatively rather than independently.

4

The Dynamic Relationship between Research Universities and the Development of High Tech Industry

Karl S. Pister

At a congressional committee hearing held in California's Silicon Valley in 1984, Jerry Sanders, president and chief executive officer of Advanced Micro Devices, said: "What's the secret of our success? In a word, people." Robert N. Noyce, vice-chairman of Intel Corporation, stated: "What 'Silicon Valley' companies have in common is that they are knowledge intensive, develop and use new technologies, and as a result, often compete in markets that are still emerging" (U.S. Congress, 1984). Noyce also listed "4 M's" that are crucial to the success of Silicon Valley types of companies: "Money, Manpower, Markets, and Motivation." These brief quotations particularly attest to the role of highly trained and often entrepreneurial people in the development of high technology industries and, thus, to the prominent role of higher education.

The success of these industries is tied directly to the mission of research universities in the United States—namely, the discovery and dissemination of knowledge through teaching, research, and public service. Noyce's "4 M's" emphasize the industry–university connection in a context that includes government, because money, markets, and motivation are inevitably linked to political, social, and economic policies. In regard to manpower, although computers are the engines of the information age, people are still their masters, and the world of high tech is the arena where the action is.

To understand the role of universities in the development of high tech, one must keep in mind that a productive and profitable industry–university relationship rests on the continuous supply of new knowledge and well-educated engineers and scientists with an entrepreneurial spirit, together with grass roots contact between industry and university researchers. Accordingly, this chapter will begin with a simple system model depicting the interaction of universities and high technology industry, emphasizing the relationships among private industry, universities, the federal government, and state governments. I will then give examples of the kinds of interactions that take place (or should take place) among these institutions to foster vigorous high technology industries. (These examples will be limited to the United States and the state of California and will reflect my personal experience at the University of California at Berkeley.) In addition, I will briefly touch upon some issues of supply and demand for engineering graduates that have significance in assessing the strength and growth potential for high tech industries in the United States.

Although the emphasis in this chapter will be on the role of universities in fostering the success of high tech industries, there is an interesting feedback phenomenon that should be noted. Skilled engineering manpower is the sine qua non of competitiveness in industry in general, but especially in high tech industries. In turn, the success of high tech industries creates a powerful magnet that not only attracts an increasing percentage of university freshmen but also attracts the group with the highest scholastic achievement at entrance. The number of engineering bachelor's degrees awarded by U.S. institutions has more than doubled in the past decade. The growth rate of degrees in electrical and electronic engineering together has been 4.3 percent, and in computer engineering the growth rate has been an astounding 20.9 percent. Typically, freshman entrants in engineering have the highest grade point averages from high school and the highest SAT scores among all freshmen. Although enrollment trends at the graduate level have followed a very different pattern, the overall impact on institutions has been substantial, but by no means positive in all instances.

A recent paper by John S. Mayo (1985) emphasizes attributes of high tech industries that highlight their close connection to universities. Academic research and industrial R&D are the sources of the technological force that drives the flow of innovations into productive use in society. Mayo calls attention to the technology gate through which high technology must first pass (figure 4–1). It is the business of universities to expand the limits of technology through basic research and to provide the flow of skilled researchers to carry out the R&D necessary for product development and manufacturing.

Education also has a secondary and increasingly important role in controlling the flow of innovations into society. Public receptivity to the products of high technology as well as governmental regulation and legislation constraining their development, sale, and operation are becoming an increasingly complex and controversial arena in our society. A technologically literate public, and technologically literate politicians are essential for the intelligent use of high technology in society.

Let me turn now to the research university and its connection with high technology industry in the United States, recalling that Noyce's "4 M's" indicate the desirability of embedding the discussion in a system containing industry,

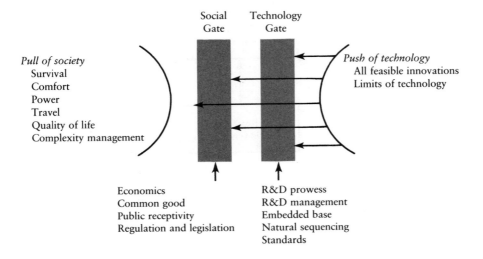

Source: Reprinted from *Information Technologies and Social Transformation*, 1985, with permission of the National Academy Press, Washington, D.C.

Figure 4–1. The Flow of Innovations into Society

universities, and government. Figure 4–2 shows a model of the interactions to be discussed in this chapter.

The Research University and High Technology Industry

Research Universities in the United States

To understand how higher education and high technology industry interact, it is necessary to touch briefly on the higher education universe in the United States. Virtually all academic R&D is carried out in the 184 doctorate-granting universities. The top hundred of these (herein called *research universities*) receive 85 percent of all federal R&D funds, and the top ten about 25 percent. The remaining institutions of higher education (about 2,900) are vital to the health of R&D universities and the nation because they educate 75 percent of all undergraduate students, although virtually no research is centered there.

Research universities, whether public or privately supported, share common characteristics that define and constrain their relationships with government and

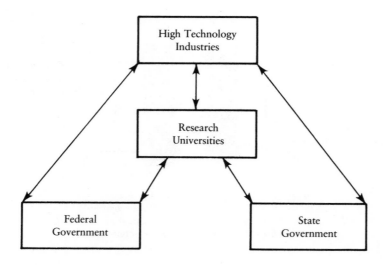

Figure 4–2. Education and High Technology Industry: A Model of the System in the United States

industry: (1) teaching, research, and public service are closely related and often inseparable; (2) research is a personal, creative process for faculty and graduate students; (3) faculty members require and enjoy considerable independence and autonomy; (4) freedom of intellectual inquiry is essential to the research environment; (5) basic (long-term) research is the principal activity; (6) applied research (built on basic research findings) is carried out in professional fields such as engineering, business, and medicine; (7) federal support of basic research predominates over other sources; (8) institutions may stimulate certain areas of research in response to state or national needs; (9) research priorities are set by faculty and are not amenable to long-range planning; and (10) peer review of proposals is the basis for selection for funding.

Given these characteristics, one may wonder how it is possible for universities to significantly affect the development of high technology industries. To understand this, we must look at the interactions depicted in the system model shown in figure 4–2. The federal government dominates the support of university research. Although the total support of university research by private industry is small, its concentration of support in engineering and the physical sciences makes it an increasingly important factor for high technology growth. Furthermore, the rate of growth of industrial support of university research has been strong in recent years.

Funds for Research: Government Support

Federal funds for research flow primarily from the National Institutes of Health, the National Science Foundation, the Department of Energy, and agencies of the Department of Defense. The linking of mission-oriented agencies such as those in the Department of Defense takes place in both basic and applied research, whereas the National Science Foundation, as its name implies, is primarily in the business of supporting basic research. The nature of this support ranges from grants supporting the research of individuals or small teams of investigators to continuing contracts for the operation of large facilities and broad institutional support. Key features of individual grants or contract support include (1) salaries of graduate and postgraduate research assistants; (2) partial salary support of faculty investigators; (3) funds for purchase of special equipment; (4) travel and publication costs; and (5) provisions for recovery of institutional overhead.

The National Science Foundation recently initiated a new program designed to develop fundamental knowledge in engineering fields that will enhance the international competitiveness of U.S. industry and prepare engineers to contribute through better engineering practice. This program establishes centers for cross-disciplinary research in engineering by providing funding for a five-year period.

The newly established centers are to provide for working relations between students and faculty, on the one hand, and industry engineers and scientists, on the other. The programs are intended to emphasize the synthesis of engineering knowledge and to integrate different disciplines in order to bring together the requisite knowledge, methodologies, and tools to solve issues important to engineering practitioners. The programs are also intended to contribute to the increased effectiveness of all levels of engineering education. They are to be located at academic research institutions in order to promote strong links between research and education. Each center is to focus on a particular subject area of national importance, such as systems for data and communications, computer-integrated manufacturing, computer graphics design, biotechnology processing, materials processing, transportation, and construction. Six centers were established in 1985, an additional five were announced in 1986, and others are contemplated as the program continues.

The president's Young Investigator Awards Program, sponsored by the National Science Foundation and the Office of Science and Technology Policy of the executive branch of the federal government, is another example of efforts by the federal government and industry to improve the ability of universities to meet industry's demand for highly qualified scientific and engineering personnel. During each of the first two years of the program, a group of 200 young engineers and scientists was selected on the basis of merit and their potential contributions to the nation's scientific and engineering effort. Persons selected must hold a faculty appointment at a university in the United States. Each awardee receives a minimum of $25,000 of federal funds per year. To further strengthen the ties between the academic and industrial sectors, up to $37,500 of additional funds per year are provided, on a dollar-for-dollar matching basis, by the NSF contributions. Thus, $100,000 in annual support is potentially available for a period of five years. When one adds to such examples of federal support of R&D in research universities the federally insured educational loan program available to students, it is evident that graduate education in the United States today, whether or not connected with high technology areas, is very dependent on the federal administration.

Although this by no means exhausts the scope of federal interaction with universities, I will turn to a second kind of governmental interaction—interaction with state governments. Although the interactions between the federal government and both public and private educational institutions are quite similar, interaction with state government is different. Berkeley and Stanford, for example, are comparable institutions in many respects: both have played critical roles in fostering the growth and development of high technology industry in California, and in the Silicon Valley in particular, and both depend on federal support of their research programs. However, whereas Stanford depends on substantial student fees for its operating budget and enjoys relatively modest state assistance through the California State Scholarship Program, many of Berkeley's substantial teaching and research facilities, as well as its operating budget, are subsidized by the state. This core support is necessary but not sufficient to carry out the educational mission of the Berkeley campus. (Currently, about half of the annual operating budget of the College of Engineering at Berkeley comes from the state.)

I have already cited the important role of the federal government in supporting research and graduate education. In engineering, the role of private industry has become invaluable in supporting research and providing state-of-the-art equipment. In addition, at Berkeley, private industry has provided capital funds for construction of a new facility for research in microelectronics and computer-aided design. This cooperative effort will be discussed later as part of the discussion of university–industry relations.

An excellent example of university–industry–state collaboration is found in the Microelectronics Innovation and Computer Research Opportunities (MICRO) program. The objective of this program, established by the state of California in 1981, is to help the California electronics and computer industries maintain their competitive edge by expanding pertinent research and graduate student education at the University of California. Under the research part of the program, faculty members submit proposals for research that potentially will be the basis for new industrial products some years in the future. In 1984–85, the program involved (1) eighty-nine projects on six University of California campuses; (2) $3,408,000 in state support funds; (3) $8,008,600 in industrial contributions (fifty-eight companies); and (4) $405,000 in graduate student fellowships. The state and industry jointly support projects that are selected by peer review. Each faculty member is responsible for obtaining a prior commitment from an industrial sponsor to support at least half of the cost of the project. Graduate student education is supported both through research assistantships funded by the projects and through fellowships granted directly to students in fields related to MICRO. Overall guidance of the program is in the hands of a policy board composed of three representatives each from industry, state government, and the university.

Funds for Research: Industry Initiatives

In recent years, a number of actions have been initiated by industry to strengthen engineering and science teaching and research in the United States. A wide spectrum of modes of support has developed, from substantial donations of state-of-the-art equipment or offers of purchase at large discounts, to direct cash gifts to institutions for such purposes as upgrading the educational environment and establishing fellowships. The following are two examples of such initiatives, both in the electronics area.

The American Electronics Association (AEA). This voluntary trade organization, comprising some 2,800 member companies, has the following activities as its major agenda:

> Establishing an AEA Electronics Education Foundation to stimulate the flow of company resources to universities. The money supports fellowships and loans for U.S. citizens to get doctorates and teach engineering, faculty grants, and equipment.
>
> Setting a standard for each company to provide 2 percent of its R&D budget to engineering education.
>
> Establishing industry committees in states and regions to raise funds and to work with state legislatures and universities to improve technical education budgets and programs.
>
> Working on federal and state technical education legislation, primarily supporting those that encourage partnerships among industry, education, and government through tax incentives and other jointly leveraged measures.

The AEA has been very effective in carrying out its objectives in terms of its direct support of university programs through fellowships and loans that assist prospective faculty; its encouraging gifts of equipment; and its effective lobbying for legislation to encourage industry–university partnerships through tax incentives. The AEA provides an excellent model for other industries to follow in strengthening the collaboration among industry, universities, and government.

The Semiconductor Research Corporation (SRC). This is a nonprofit, tax-exempt organization whose forty member companies are of diverse sizes, but in one manner or another are involved with the manufacture or utilization of semiconductor equipment or materials. The essential purpose of SRC is to promote basic research and scientific study by universities in engineering, mathematics, and the physical sciences underlying semiconductor technology. SRC operates by pooling membership fees and contracting with universities for the creation of research centers as well as for specific research projects. The 1984–85 annual budget for research projects totaled $12 million, which was disbursed among thirty-seven universities and involved more than 180 faculty members and 400 graduate students.

University–Industry Interaction

I have already given examples of ways in which the high tech industry has taken collective action to promote its continued growth in the United States by strengthening university education. Taking the University of California at Berkeley as an example, we can look at this relationship in greater detail. I should set the stage by recalling that highly skilled manpower is a central characteristic of high technology industries.

The contribution of research universities to the success of high technology industries through teaching, research, and public service activities is made both through the creation and dissemination of new knowledge and through the training of people whose education and research experience prepare them for high tech employment. Indeed, probably the major motivation for industrial support of universities is the recognition of the desirability (and necessity) of maintaining a steady flow of new engineering talent. For example, it is not a coincidence that Silicon Valley companies have made major contributions to and investments in Stanford University and the University of California at Berkeley. For the past several years, the state of California has awarded more engineering degrees than any other state. More significantly, Berkeley and Stanford together awarded 39 percent of the engineering master's degrees and 58 percent of the engineering doctorates in the state in 1984. At the dedication of the Berkeley CAD/CAM (computer-aided design and manufacturing) facility in September 1985, Advanced Micro Devices president Jerry Sanders made this observation that illustrates the point:

> And there is an even more basic payoff for industry, because the university is the source of the engineers and scientists we need to keep on providing innovative

solutions to electronic design problems. Since 1979, AMD has hired 97 people from Berkeley, including a dozen Ph.D.'s. We want more than our fair share of your talent.

Emphasizing the critical need for close collaboration between universities and industry, he went on:

> But industry can't do it alone, and in fact it shouldn't even try. Technology, after all, is the application of science to society, and in order to maintain technical progress we need a close relationship with the people who do basic scientific research. That means close cooperation between industry and the university. We need research into new processes, new materials, new design techniques, new architectures, even new cross-disciplinary curricula combining manufacturing, electrical, and mechanical engineering. That is why the Berkeley CAD/CAM consortium . . . is so important.

Research and teaching activity at Berkeley have been enhanced by the renovation of space for a state-of-the-art microfabrication facility for integrated circuits and the addition of space and facilities that permit 200 graduate students to carry on research in microelectronics and CAD. What makes this program unique is that although the state of California provided funds for the microfabrication facility, the new facility was made possible by an industry consortium of fifteen member companies that pledged $8.5 million to construct the new research center and more than $10 million for computer-related equipment. As a state-supported institution, Berkeley enjoys such private support because the College of Engineering, and particularly the Department of Electrical Engineering and Computer Sciences, historically has maintained close grass roots connections with industry. Faculty and graduate students maintain close contact with their colleagues in industry through a system of interchanges. Faculty members receive industry-financed research grants and may be employed by industry for the summer or as consultants. Similarly, industry provides students with important training and experience through summer employment.

The Industrial Liaison Program of the College of Engineering at Berkeley provides a framework for companies of any size to interact with engineering faculty and students. A company with interests in several departments or research programs can affiliate by making a nominal annual contribution in unrestricted funds to the college. Companies can also choose to affiliate with a single department or program for a lesser annual contribution. Currently, more than 250 member companies are affiliated with this program. An annual conference featuring short presentations on current research programs in the college attracts some 500 attendees.

The Engineering Cooperative Work-Study Program (CO-OP) offers students the opportunity to be employed in industry and government while completing their education. CO-OP students alternate between classwork and practical employment experience by spending six months of the year in school and six months on the job. This university–industry partnership creates a flexible and continuing link between theory and practice at an early stage of professional training, and a high percentage of CO-OP students become permanent employees upon graduation. Currently, about 10 percent of undergraduate students beyond the freshman year are working in more than 150 companies.

What Makes the System Work

Thus far, I have described some of the interactions that can and should take place among the institutions represented by the model. If the purpose of this system is to develop and sustain growth and profitability of high tech industries in the United States, careful tuning of the interactions is needed for the system to perform optimally—and it is not possible to consider this system apart from the U.S. economy as a whole or apart from political considerations. The following illustrations help explain the complexity of the problem.

Tax Laws Affecting Industry

During the current federal administrative term, changes in the tax laws on capital gains have reduced the effective tax rate substantially (from 49 percent to 20 percent). The result has been a significant increase in the amount of new money committed to venture capital pools—the lifeblood of the kind of start-up companies that characterize Silicon Valley's history of success. Similarly, the R&D Tax Credit Act reduced the tax burden on retained earnings placed into R&D efforts, thus encouraging expenditure increases for R&D internally; from the standpoint of universities, it has created a stimulus for gifts of both equipment and cash to support university research at a time when there is a very urgent need for such assistance. Discussions now taking place in Washington concerning the revisions of tax laws place the future of these important changes very much in doubt.

Freedom of Inquiry and Publication

When industry and university collaboration intensifies, as it has done markedly in the last decade, there arises a justifiable concern about the extent to which control of research directions passes from the hands of faculty to those of industry. Although it is generally agreed that the line between basic research and applied research in engineering is often difficult to draw, there is nevertheless danger that industrial pressure may push university researchers too far toward applied research because of their willingness (or need) to accept industrial support funds. The Semiconductor Research Corporation program discussed earlier is an excellent model that preserves individual freedom of inquiry yet serves the needs of the industry for basic research.

A related pressure may arise in the area of publication rights. Experience at most research universities supports the advisability of allowing no restrictions on open publication of research, irrespective of funding source.

Patent and Intellectual Property Rights

The enthusiasm for industry–university collaboration is often dimmed by the legal entanglements that arise when ownership of research-generated devices, processes, or software is examined by the lawyers for the respective partners. Currently, there is a rather wide range in content of agreements that cover this area. The issue of ownership of software has surfaced fairly recently and has given rise to extensive discussions on the rights to intellectual property. An oversimplified description of the current protection status of software is that it is between copyright and patent registration. Clearly, this issue will gain in importance as more marketable software is written.

Regulation of Industry

The biotechnology industry is an example of an emerging technology that will (or has already) run into the regulation roadblock. No one will argue that genetic engineering, which seeks to affect cell behavior in order to produce living organisms for industrial processes, poses risks for society that are similar in many respects to those posed by nuclear engineering. The response to identifying risks in society inevitably is governmental regulation. I cannot debate the proper balance between freedom and regulation here but wish only to note the following: The San Francisco Bay Area has become a major world center for the biotechnology industry. More than twenty companies, some 10 percent of the U.S. total, are located there. (It is no coincidence that four major university research centers engaged in genetic research are located there also.) The response to this concentration of biotechnological activity is a labyrinth of federal, state, and Bay Area agencies whose decisions will be instrumental in shaping the future of this industry in the United States (Kureczka, 1984). The bottom line in this kind of technological balance sheet was captured in a lecture delivered by Simon Ramo (1984), co-founder of TRW, Inc.:

> It is not that lawyers, businessmen, politicians, bankers, accountants, and other groups should participate less in the development of U.S. technology policy, it is rather that engineers should participate more.

In summary, it is most unlikely that any one model of university–industry–government interaction will suffice. The pluralistic character of research universities, differences in social and economic conditions in the various states, and shifts in federal policy virtually guarantee this result. However, it is equally clear that one must retain the kind of system overview depicted in figure 4–2 when contemplating ways and means of effecting change in high technology industries.

Issues for the Future

Looking to the future and the factors or conditions that are likely to affect the ability of high tech industry to continue to grow in the United States, one can expect the following to predominate: (1) emergence of a truly global economy, placing added pressure on U.S. industrial competitiveness; (2) adequacy of the pool of engineering faculty; (3) adequacy of the resource base for engineering schools; and (4) a need for widespread public awareness of technology.

It is hardly necessary to call attention to the pace at which U.S. industrial competitiveness has eroded in a number of areas central to the U.S. economy. At a recent symposium, "U.S. Industrial Competitiveness and the National Academy of Engineering," N. Bruce Hannay, retired vice-president for research and patents, Bell Laboratories, and chairman of the National Academy of Engineering/National Research Council Committee responsible for a series of studies of U.S. industry and its competitiveness in world markets, noted the following common theme characterizing the group of industries studied:

> Despite the disparate nature of these industries, all must now be termed world-scale industries. They must be managed in that context, and U.S. public policy must reflect the reality of growing and more pervasive international competition. That this has not yet been achieved is apparent in a number of ways: the low level of proficiency in foreign languages and limited knowledge of foreign cultures provided by our educational system; government policies that reflect concern only with domestic issues, with little regard for their effect on our international competitiveness; the recurring hostility between government and industry on market matters and the bureaucratic tangles that ensnarl licensing, certification, and approvals; the control of technology transfer in ways that are inconsistent and lack perspective on the availability to foreign customers of alternative sources; the comparatively low emphasis put on international experience in U.S. executive development programs; and the comparative disadvantage of U.S. firms with respect to public financial and tax supports for foreign trade and for the capital investments needed to keep us competitive.

Reflecting the same concerns, a 1985 National Research Council report on engineering education and practice in the United States concluded:

> In the context of an increasingly global economy, American engineers must become more sensitive to cultural and regional differences, so that they can

design products that foreign markets require and will accept. Engineers will also need to appreciate the financial, political, and security forces at play internationally. The nontechnical components of engineering education ought to include exposure to these aspects of contemporary engineering. In addition, the engineering community should strive to ensure open communication on these matters among engineers and companies the world over. (p. 12)

Application of these observations to the redesign of engineering curricula presents a major challenge for U.S. engineering institutions. The sentiments expressed in these quotations naturally presuppose the continued excellence of the technical content of university engineering education and the continued ability to provide highly skilled engineers for U.S. industry. Whether or not the traditional four-year curriculum will remain as the norm is a challenging question for U.S. institutions and industry.

A second major factor that will affect the ability of engineering schools to carry out their mission is the probable shortage of qualified engineering faculty over the next ten to twenty years. At the beginning of the current decade, U.S. engineering schools experienced a 10 percent vacancy rate for full-time teaching positions; that rate has improved slightly over the ensuing years. However, in selected fields, such as electrical and computer engineering, the situation remains critical in many institutions, where student enrollments have been allowed to grow in response to unprecedented student demand for admissions to engineering programs. The major cause of the vacancy rate is the inability to attract American Ph.D. recipients to academia. Not only do the "best and brightest" avoid faculty careers, but the proportion of all engineering Ph.D.'s who elect not to teach is increasing. A study comparing the 1982 Tau Beta Pi engineering honor society class with that of 1977 found "a profound lack of interest in teaching" and "little interest in full-time engineering graduate study" (American Electronics Association, 1984, v, vi). Were it not for a dramatic increase in the number of Ph.D.'s awarded to foreign nationals, U.S. engineering schools could not have met the dramatic increase in enrollment that has taken place. Eighteen percent of all U.S. engineering faculty in the fall of 1982 were foreign-born, and more than one-fourth of all assistant professors of engineering had not received their B.S. degrees from U.S. institutions (American Electronics Association, 1984). These foreign nationals generally have not displaced U.S. citizens, who typically tend to leave after the bachelor's or master's degree.

These observations are consistent with the study recently completed by the California Postsecondary Education Commission (1985), which noted:

The high percentage of doctorates in Engineering and Computer Science awarded to foreign students both nationally and in California appears to result from the "siphoning off by industry" as one dean put it, of American students after they have completed baccalaureate or master's degrees. . . .

Judging from quite sketchy statistical evidence, it appears that perhaps as many as half to three-fourths of the foreign students who earn graduate degrees in California are employed in this State and Country after graduation. . . . (p. 28)

Nor will the problem improve in the decades ahead. Institutions in the United States will face massive faculty retirements in the 1985–2005 period because of the rapid expansion of the faculty ranks following World War II. Although there have been a few relatively modest attempts to address the faculty shortage problem—by U.S. industry and federal agencies, particularly the Department of Defense—there is little evidence of a national plan to deal with this vital problem.

A companion problem to the faculty shortage that continues to plague engineering institutions is that of resources for instruction and research. Although industry takes it for granted that competitiveness demands investment in physical plants and facilities, application of this self-evident principle in the academic world is less than satisfactory from the standpoint of U.S. engineering schools. Recent estimates indicate that a national investment of more than $1 billion would be required to bring physical facilities located in engineering schools in the United States up to satisfactory standards. Compounding the problem is the enormous shortfall of funds to replace and maintain laboratory equipment needed for research and instruction—estimated by some to be another $1 billion problem. Industry's awareness of the state of engineering laboratory equipment, coupled with favorable tax laws, has led to very substantial contributions of equipment to engineering schools. Reference has already been made to one such instance at U.C. Berkeley: the microfabrication facility for research in microelectronics and CAD. Unfortunately, as many institutions are learning, maintenance costs for computers and other electronic equipment typically average 10 to 12 percent of the purchase price per year. Industrial gifts of equipment seldom include provision for maintenance, and many institutions are facing a situation in which they are forced to turn down offers of gifts of needed equipment because they lack the resources for maintenance. Institutional resources have simply not kept pace with the demands growing out of the increasing complexity of engineering laboratory equipment, and no solution is in sight for this problem.

Finally, no discussion of education and technology can be considered complete if it neglects the public's role and the need for public understanding of both the promises and the problems of technology. Mention has been made of this in connection with the model for flow of innovations into society depicted in figure 4–1. Public opinion, as manifested through the politics of regulation and legislation, can be decisive in determining the limits of high tech growth. The emerging industries based on genetic engineering and biotechnology are prime examples of industry dependence on the public's level of technological literacy.

Without an informed public, a recurrence of the history of the nuclear power industry is likely. At a recent conference, MIT president Paul Gray highlighted the importance of technological literacy by quoting an MIT colleague, Philip Morrison:

> If we cannot promote the growth of wider understanding of the world view of science and technology, we endanger not only our own abstract enterprises, but even the essence of democracy. For the necessities of economics will eventually enforce a social division into islands of the trained who understand enough to devise and operate an increasingly complex technology, with a sea of onlookers, bemused, indifferent, and even hostile.

At stake here is less a problem for engineering schools than for the institutions to which they are affiliated; it seems obvious that the meaning of a liberal education in the information age is a question that cannot be ignored.

Part III
Human Resource Management Concepts and Practices

The chapters in this part are concerned with the contemporary human resource management approaches linked to high technology; understanding the origin and rationale for innovation in human resource management; and placing innovations in historical and economic perspective. Among the specific issues analyzed are human resource management practices that relate to recruitment of personnel, maintenance of skills, and employee commitment and compensation systems in light of changing organizational characteristics and business strategies over the life cycle of the firm.

In chapter 5, Fred Foulkes approaches these topics through a case history of AutoTel Inc., a successful high tech firm. The case highlights the challenges to human resource management as the firm moves from a start-up to a highly successful enterprise. The history of this firm demonstrates how human resource management practices are called into question as a firm changes and matures. Although these experiences are in many respects unique to AutoTel, they also illustrate general issues that affect the ways in which the management of human resources in most high tech firms is related to the business fortunes of the company.

A firm's own growth pattern can present a major challenge. As firms mature, the need for formalized human resource management procedures and some degree of bureaucratization often goes unremarked, since it is so natural a consequence of growth. Yet there are many options available and difficult decisions to be made as this occurs. AutoTel's experience highlights this fact. The company's early history was characterized by rapid financial and employment growth. Now, the issue of the role of human resource management, with its link to the productivity of the professional work force, may be the crucial issue at AutoTel.

This case history poses an old question: Is it possible for companies to plan ahead in regard to human resource management? The extent to which a firm can adopt a particular approach, institute policies and procedures, and be confident that it has done the right thing is not always clear. Until recently, AutoTel's mode has been reactive; for the most part, human resource management decisions and policies were made on an ad hoc basis, without being explicitly linked to the business strategies of the firm.

In chapter 6, George Milkovich looks at an area of human resource management—the design and management of compensation systems—in which innovation is believed to be particularly common in high tech firms. He examines three relevant areas: (1) basic compensation policies, (2) whether such policies in high technology firms differ from policies of other firms, and (3) the strategic perspective in which compensation policies are set. He reviews the state of knowledge as well as some of the fads and fashions in high tech compensation.

Milkovich considers four policy decisions basic to the design of pay systems: the mix of pay forms, positioning a firm's pay relative to the pay of its competitors, the pay relationships among jobs and skill levels within the firm, and performance emphasis. In each of these areas, firms have a choice of whether to depart from traditional compensation practices in order to implement more innovative practices.

According to Milkovich, the innovative practices attributed to high tech are reminiscent of the early pay practices of firms that are now more mature, calling into question the extent to which such practices are a result of something new and different about high tech. They may result from efforts like those of many other newly established, rapidly growing firms to prosper and survive—regardless of the nature of the product. Milkovich also cautions against regarding high tech firms as a homogeneous group. He suggests that the variability among firms in the industry illustrates the need to identify factors that influence choice of compensation policies.

Chapter 7, by Robert Miljus and Rebecca Smith, is concerned with the changing role of the human resource professional as the firm changes and as it responds to other factors that affect its human resources and business activity. The authors emphasize that the tasks of human resource management, which include the effective acquisition, compensation, development, and utilization of the firm's human resources—with special stress on scientific and technical workers—are of paramount importance to its performance and its ability to compete successfully in the global marketplace.

Readers will be especially interested in Miljus and Smith's characterization of adaptive organizational conditions—the conditions best calculated to ensure that a firm can adapt to demands from the economic and political environments as well as to its own stage of development. In particular, they address the human resource manager's role in that adaptation.

5
Human Resources at AutoTel, Inc.

Fred K. Foulkes

> The history of the human resources function at AutoTel has been a very rocky one. The management team does not think much in terms of personnel policies and practices, yet today there is increasing focus on human resources and what they can accomplish.
> —Charles Long, Chief Financial Officer

In the spring of 1985, Charles Long, chief financial officer, contemplated the meteoric growth of AutoTel and its impact on human resources. A producer of automatic call distributors (ACDs) for telemarketing applications, Auto-Tel (founded in 1971) had the largest installed base of standalone ACD systems. Yet only a few years earlier, AutoTel had been a small company in which all employees and officers worked under the same roof and shared snacks, conversation, and company decisions in the hallways. Since those early days, the human resources staff had been occupied primarily with recruiting in order to fill the company's constant need for new employees. By the spring of 1985, the company employed more than 1,800 people, 50 percent of whom had been with the company less than eighteen months (see table 5–1 for employee growth statistics).

The assimilation of so many new people over such a short period of time, plus an anticipated deceleration of AutoTel's phenomenal growth rate, had led to a call from all parts of the organization for more attention to human resource issues. Concern centered on how AutoTel was going to install more formal personnel procedures yet maintain its small-company, nonbureaucratic, and entrepreneurial character. One nine-year veteran wondered: "How are we going to formalize our informality?"

As the senior management person to whom the newly appointed personnel director would report, Long reviewed the history of the company. He considered the unique culture of AutoTel, and tried to determine the strengths and weaknesses of the company's management of human resources.

The author acknowledges and appreciates the assistance of Graham Holmes in conducting research for the case study and preparing the original version.

Table 5–1
Growth at AutoTel, 1977–84

	1977	1978	1979	1980	1981	1982	1983	1984
Employees	150	210	286	389	563	895	1229	1773
Revenues ($M)	10.2	13.4	19.5	32.4	43.2	86.4	132.0	202.5
PAT[a] ($M)	1.1	2.0	2.6	5.0	8.4	12.7	18.1	27.2
EPS[b]	.07	.10	.12	.20	.29	.44	.59	.89
R&D ($M)	1.0	1.5	2.2	3.3	6.7	11.0	17.7	27.1
Working capital	1.9	6.1	6.3	26.1	28.2	52.2	58.1	69.6
Cash and equivalents	.07	.36	.97	24.2	21.5	48.4	52.8	62.6

[a]Profit after taxes.
[b]Earnings per share.

The AutoTel Way

The development of a distinct AutoTel style had paralleled the company's rapid growth. It was largely derived from the philosophy and values of Bradford Bennett, the founder, chairman, and chief executive officer. Unlike most CEOs in the world of high tech executives, where engineering degrees and conservative politics were the rule, Bennett had earned an undergraduate degree in business and was a liberal Democrat. An older executive in the high tech industry at age of 55, his strength was marketing and his style was somewhat conservative, personal, and low-key. One employee remarked about Bennett: "You know that he cares; he's real down to earth and easy to talk to." Observers acknowledged that among his strengths, Bennett's ability to let go of responsibility at the right time had been important to the successful development of the company.

"There is a real style here . . . taking a chance on people," one manager in the development department explained. "The way things get done at AutoTel is that the right person for a job is found and then given practically free reign. There are high professional expectations, mutual trust, few rules, and no time clocks." Both Bennett and Bill Price, the president and chief operating officer, were credited by employees with making this strategy work—not only by recruiting outstanding people but also by not being afraid to fire the misfits. One manager commented: "Their example has impact throughout the company. When you know something is right to do, then you do it." They encouraged people to present their views and opinions. "Be honest—if something bothers you, say it. Don't be a 'yes man' . . . don't overanalyze at the expense of action."

The individual who "made it" at AutoTel (the "AutoTel type") was described by employees as being a self-starter, assertive and energetic ("capable of more than 100 percent"), with a "get-it-done" personality. Above all, the AutoTel type was a team player with good judgment and common sense. When new employees were initiated into the AutoTel way, they learned that the company's nature was cooperative: departments worked together. There were few turf issues; instead, there was a strongly shared commitment to the success of the company. As a result, internal pressures could become intense at the close of each quarter or project. "It's a fast-paced environment," one senior manager pointed out. "You have to be able to juggle a lot of things. And you are going to drop balls sometimes. You just try to make sure they're the right balls."

Although employees were pushed and pushed themselves hard, they were also well taken care of. In the words of one employee: "When the news is always good and growth translates directly into opportunity, it is easy to be enthusiastic, confident, and committed." A "can-do" atmosphere was the mark of the company's entrepreneurial heritage, and the consensus among employees was that it had played an important part in AutoTel's success.

The ACD Industry

As a result of continual increases in the price of the average sales call and because consumer behavior favors convenience products and services, the use of telemarketing as a sales medium has increased considerably over the past decade. Responding to this growth, a number of telecommunications companies have developed automatic call distributors (ACDs) to increase the efficiency of handling a large number of incoming "800" service-type calls.

The primary function of an ACD is to maximize the number of calls answered during a given period of time by routing calls in such a manner that the number of free agents is minimized. For example, most airlines use ACDs to handle the large volume of incoming calls. Numerous studies have indicated that passenger volume is significantly higher for the airlines that are most successful at expediting the reservation process. In other words, if a passenger cannot get through to an airline, or is continually put on hold, another airline will often be selected. These findings have prompted a number of large airlines to evaluate their reservation centers carefully, and to install the most sophisticated ACD systems. Other large users of ACDs include insurance companies, telemarketing service bureaus, and order entry catalog houses.

Telephone system vendors—including AT&T, Northern Telecom, and Rolm—have traditionally been the leaders in the ACD market, offering software packages that are added to a company's existing private branch exchange (PBX) phone system. However, because of the expanded interest in telemarketing,

AutoTel and others introduced standalone ACD systems that offered benefits not associated with PBX-based systems. These benefits included greater flexibility in system size and increased cost savings resulting from improved call-routing capabilities.

AutoTel was followed into the standalone ACD market by two larger companies: Rockwell International and Teknekron Infoswitch. The increased competition in the marketplace has posed a number of challenges for AutoTel. First, increased research and development from the PBX vendors have increased the functionality of the PBX-based systems, thus reducing the number of advantages enjoyed by the standalone system companies. In addition, the large installed base of AT&T, Rolm, and Northern Telecom PBXs provided a much easier target to which ACD software packages could be marketed. This fact, combined with the increase in R&D spending, also reduced the number of advantages enjoyed by the standalone system companies over the PBX vendors.

A third challenge for vendors resulted from the growing requirements of users that imposed more demanding conditions on ACD system creators. Engineers and programmers capable of state-of-the-art development were scarce and thus costly. Moreover, an exceptionally high turnover rate for many companies, though speeding up the spread of knowledge throughout the industry, was providing little return on investment in in-house training for college graduates. Finally, the growing diversification of user applications in telemarketing warranted greater customization of ACD software, which increased the demands on both human and financial resources. These constraints, and the competition of a growing number of telecommunications giants, have been especially threatening for companies like AutoTel.

Company Background

Brad Bennett started AutoTel in 1971 with $2 million raised from thirty-five individual investors in exchange for 45 percent of the company. Bennett's mission was to provide a high-performance ACD that would compete in price with the PBX-based systems and in capability with the products of the larger standalone vendors. The company began with 2,000 square feet of rented space on the second floor of a rehabilitated factory in Mountain View, California.

The first few years were difficult. Bennett recalled that on one particular day, "We had only $2,500 in the bank and a $40,000 payroll due. A $40,000 check came in that day." A generalized ACD was the company's first viable product, but at $5,000 per agent (with an average system size of fifty agents), it was not an easy sell. Although the performance of the product was excellent, it lacked any unique feature that would allow it to stand apart from the competition.

In 1974, with revenues at $5 million, Bennett decided to change the company's direction and develop a PBX-based software product that would be compatible with any PBX. "It was a bet-your-company decision," Bennett said.

"At that point, this type of software was offered only through the very large PBX vendors, who had an established reputation within the industry and a secure customer base." Bennett planned to outsell the PBX vendors through a number of system enhancements. In addition to greater hardware compatibility, AutoTel's product would offer the most complete call-reporting features available. The product was well received by users, and in 1976, the company earned a profit for the first time. Bennett began to expand the company: he hired his first sales manager, bringing the total number of employees to eighty; offices were opened in Waltham, Massachussetts, and Dallas, Texas; and various overseas sales agents were established.

Expansion

Expansion at AutoTel occurred through both internal and external means. Internal expansion was supported by R&D expenditures at around 12 percent of sales, which was low in comparison with competitors. According to the company, this was largely due to high employee productivity. External expansion, supported in part by public offerings in 1978 and 1980, that netted $28 million, played a major role in the development of more sophisticated software packages.

The most important ongoing strategy for product development was listening to customers and responding to their needs. In the words of Brad Bennett:

> We design our products in response to user needs, not in a vacuum. We interface with our clients continuously. If they need a particular product or if a product needs a particular feature, we want to know about it. We increase programmer productivity by talking to programmers; we make systems more functional by talking to users; and we discover what management's needs are by talking to management.

AutoTel was not unduly worried about competition from larger companies. According to Bennett:

> The PBX vendors have no inherent advantage over us in the software business as they do over each other in hardware. It takes, in our view, a limited number of outstanding system programmers and engineers to develop a successful product. That law applies to these companies as it does to us. I'll stack up our people against theirs any day of the week.

In 1983, Bennett was able to proclaim:

> We used to lose four out of five competitive sales to one of the three leading PBX vendors, but recently we've been winning four and five out of five. A lot of telemarketing companies are now committed to AutoTel as being their software supplier and view the others as their hardware supplier.

One AutoTel salesman remarked: "Ninety-nine out of a hundred potential buyers will call AutoTel."

Operations

Top Management and Decision Making

During AutoTel's formative years, the company was largely run by Brad Bennett, who made all the major decisions. A board of directors was chosen in the early years on the basis of Bennett's "gut feeling." By 1985, president and chief operating officer Bill Price handled day-to-day operations, aided by a management committee composed of the eight top officers of the company.

The formation of a management committee in 1981 acknowledged the rapid growth of AutoTel and the need for management by more than one person. In Bennett's words: "Its members embodied two of the most important management ingredients, concern for the user and good judgment." The committee was chaired by Bill Price, then senior vice-president responsible for development of the company's product line.

Decision making at the senior level of the company was informal, although there was some disagreement about the degree of centralization. One middle-level manager observed: "It seems like everything has to go to Sean [Healy, executive vice-president] or Bill, and it slows things down." Another manager saw it this way:

> Decision making can be very informal and sometimes seems haphazard. There are no procedures and there are a few key people making most of the decisions. Though there's a lot of opportunity here because of the high growth and the company's willingness to give responsibility to people without much experience, there's not much delegation of decision-making responsibility. It can get frustrating.

One senior manager agreed:

> Too many things aren't being delegated. Department heads are making all the decisions. To delegate decisions effectively, to a great extent you need a framework in which people can operate. Part of that framework must include certain policies and guidelines. We need to do that.

Although many agreed that most important decisions at AutoTel were made by senior management, others felt that it was the informality of the decision-making process that led some managers to underestimate their degree of autonomy. One senior manager remarked: "The brass ring is always out there if you want it; you can go for it. I think people shy away from responsibility." Charles Long agreed that there was some confusion among managers: "People have more autonomy and decision-making authority than they think they do. Some stupid little things come up to the top."

Communications

Communication of top management decisions to the rest of the company was conducted on an informal, verbal basis through departmental and field managers. The company tried to avoid memos whenever possible, particularly for the communication of important issues. The danger, according to one senior manager, was the inevitable "weak links" in the communication process. "There is nothing more devastating than a manager who can't communicate accurately what he hears from his bosses. It's tougher communicating the farther you get from headquarters." A recent example of the failure of communications occurred when the company adjusted its growth goals for the second quarter of 1985 from 50 percent to 30–40 percent. The company had grown at 50 percent plus for twenty-five quarters, but a recent economic downturn that led to a rash of unprecedented layoffs across the high tech industry caused management to be concerned. Long explained that the "communication of this message, along with the necessary adjustment of expenses—which, for us, means hiring fewer people—was distorted in a few pockets around the country. Now we're having to patch up the damage as well as find out where communication broke down."

> One senior manager, who was new to the company observed: I've worked elsewhere and experienced these ups and downs; it reinforces my confidence in management to see ahead and make adjustments. And look, we're still hiring. The company is still growing. But for a lot of people here, they've only experienced AutoTel and 50 percent growth and they probably have a feeling of invulnerability—that there would be no end to it.

Middle Management

The growth of middle management paralleled the company's overall growth. By 1985, there were 380 managers, including group and project leaders. An important part of the manager's job was to develop successors from within the support staff. In the words of one technical support manager: "Today's technician is tomorrow's manager at AutoTel."

Bennett told his team that the most important guideline for identifying management potential was "good judgment and common sense. Don't get bogged down by factors such as age or experience." As a result, the company was not afraid to give management responsibility to individuals who had no managerial experience. The success of the promotion-from-within policy at the line level varied. Some employees had had good experiences with their managers, others were less enthusiastic. One employee commented: "They care about people here, but the approach is not professional. You have to look out for yourself."

By 1985, the company was becoming conscious of the need for some formal management training to offset some of the unevenness in management quality. At the time, management training consisted only of a short training program on interviewing and conducting performance reviews. These classes had been well received by some managers but resented by others.

In addition, the company was reevaluating its promotion-from-within policy, recognizing the value of recruiting from the outside, particularly at the senior levels. A senior manager noted:

> You can't always develop internally, particularly in a company growing as fast as we are. It doesn't mean an end to internal promotion, by any means, but more of a balance. We need people with larger company experience. But we will have to be careful about who is brought in—not people who want to change AutoTel, but people who will fit in and become a part of AutoTel.

Once promoted to manager, some employees admitted that when they had assumed management responsibility they had largely proceeded on a "seat-of-the-pants" basis, with some guidance from their own supervisor. One manager in technical support, who had previously worked in a bank, was overwhelmed by the contrast "between an environment where everything is procedured to death and AutoTel, where my supervisor tells me to 'do what's right!' If you feel there is a better way to do something and it makes some sense to others, then it gets done that way without a lot of commotion."

The Development Staff: Retaining Good People

There was one group of employees at AutoTel that the company worked especially hard to please: "the people who drive our business—our engineers and software developers" (Brad Bennett). As a result of the unique and creative way AutoTel housed and treated its development staff, there was very low turnover among programmers and other development employees—in sharp contrast to the rest of the high tech industry.

In 1982, company growth led to the relocation of the development operation to a site a few hundred yards down the road from corporate headquarters. More than forty members of the development team were surveyed, and the new site was designed to meet their preferences. The resulting setup recognized and encouraged the distinctive culture that had developed. The new center included offices and open-plan work areas for up to 350 software developers, twelve conference areas, a 1,500-square-foot library, a 100-seat auditorium, and a 5,000-foot atrium. The building also housed a catered cafeteria, a racquetball court, and an exercise room offering weightlifting equipment, stationary bicycles, aerobics classes, and lockers and showers.

In contrast to those at corporate headquarters, development personnel dressed casually and kept their own hours. The software developers and engineers—who numbered more than 350 in 1985, with an average age of 30—talked about the contrast of their "small-company" atmosphere to the corporate or education offices. "What counts is the quality of your work; you do what you need to get the job done. You can be yourself. You can work here and be a real person. When things get tight, nerf balls fly and people scream and shout." Many members of the development staff had come to AutoTel from larger companies and were enthusiastic about the work atmosphere and challenges at AutoTel. There was "tremendous stability" as a result, though no real career pathing had been developed. In the words of one member of the development team:

> It doesn't get boring around here. And it is instilled in all of us that people are what make AutoTel, that everybody is important to the success of the company. Brad Bennett embodies this. He is people oriented. . . . Here in Development the effect is essentially a corporate hands-off policy.

Technical Support

Originally part of the development operations, technical support was a separate 160-person operation by 1985. Ensuring the customer's ability to use the products easily demanded continual organizational refinements. Ben Baldwin, the senior vice-president for support, solicited ideas on a regular basis through group meetings with technical-support staff. "There is a real effort to get our opinions; a respect for the lower-level point of view. Our suggestions are heard and implemented," commented one staff member. A strong sense of pride permeated the department, which boasted the highest user rating in the industry and the lowest attrition rate for a job with a traditionally high burnout rate.

Support personnel pointed to their team approach, both intradepartmental and interdepartmental, as a major reason for their high user rating. Mobility within the company (job tenure was typically between one and three years) and full management support (a good work setting, adequate staffing, equitable compensation) were factors cited for the high morale, enthusiasm, and commitment of the staff. In the words of one support representative: "We are given real, interesting challenges without getting bogged down with peripheral details. If anything, you get frustrated because you can't do enough. Most pressures are self-imposed."

Sales

The sales division was organized into ten regions, each with a manager who reported to Sean Healy. Below the regional managers were eighteen district managers, and below them were 100 account executives, mostly men in their late twenties to mid-thirties. In addition to site visits, the account executives conducted seminars as part of the company's mass marketing strategy.

Until 1985, practically all salesmen came to AutoTel with a proven track record, strong technical skills, and good training from large companies such as Hewlett-Packard, ITT, and Motorola. One notable exception was a former customer-support representative who had expressed interest in sales to Sean Healy. He was teamed with the company's top salesman for nine months, and by his second year he became the company's number one salesman.

Tough quotas were set. Commissions were based on retroactive kickers for making quota. For making above quota, commissions were even higher. At an annual worldwide sales meeting, those who achieved quota (about 75 percent) became members of the "President's Club." At the three-day business meetings of the club, the year was reviewed, future objectives were outlined, and new products were announced.

The Personnel Function

History of Personnel at AutoTel

Over the years, three different human resources directors had reported to Charles Long. By 1985, he noted that the "whole organization is starting to cry for more attention to personnel issues." Many employees viewed the personnel department as playing "a very minor role, mostly recruiting." Long commented: "A lot has been the result of senior management not supporting it, doubting whether it was really necessary, and also having the wrong people in it."

In its initial stages (1977–80), Personnel had been a glorified administrative function. Though the first human resources director had embodied the company's culture of "common sense and good judgment" and had excellent organizational skills, she knew little about human resources. In 1980, the company hired its first human resources professional, who brought the department forward but, according to Long, "perhaps got in over her head and probably didn't get much management direction."

From 1981 to the fall of 1984, the new director, Jim Welch, "did a lot to legitimize the function and bring some respect to it," according to Long. In 1984, he was promoted to divisional vice-president for administration, reporting to Bill Price; and the company hired a new, highly respected director for human resources. The results were disappointing. Charles Long noted:

She made it her first priority to write policy and procedures manuals. Not that they were not long overdue, but when you come out with, for example, a ten to twelve page document on procedures for snowstorms and work cancellation when we don't call off work—if you can't get here, you can't get here, and if you can get here, you do—then there is a problem. Her orientation was definitely not right.

As a result, Welch reassumed direct responsibility for human resources, once again reporting to Long.

Organizational Changes, 1984–85

When Long regained responsibility for the human resources function, he reorganized the department. After reading the sequel to *Up the Organization*, in which it was argued that human resources departments should be abolished and reorganized to focus only on "people activities," Long did just that. His aim was to connect the administrative, financial, and marketing activities of the human resources department to the expertise of the most closely related company functions. He assigned personnel administration and recordkeeping to the controller and the benefits group to the treasurer. He also asked that the employment group get the advice of marketing communications people. The company's advertising agency was contracted, under Long's direction, to do employee advertising as well. The reorganization "removed all the paper-pushing and number-crunching from Personnel," leaving the department with three major responsibilities: hiring, compensation, and employee relations.

The level of expertise among the existing Personnel staff was uneven. Although the staffing managers were all experienced professionals with a high tech background, other staff did not have a personnel background and had not begun at AutoTel in human resources. As a result, some functions, such as compensation, had not been fully developed. Long also felt it was time to add a training and development function to Personnel.

Shortly after this reorganization, Rachel Pacquette, who had worked in customer support and education since joining the company, was appointed manager of employee relations and college relations. In addition, Ann Cowen was hired from outside the company to be manager of training and development. Her mandate was to begin a management training program. Finally, in early 1985, Welch left AutoTel and Leslie Burbank was hired from outside to be the new director of human resources. In the words of Burbank:

I gave up a business partnership at my previous company that I worked hard to get. The question for me was whether I could come to AutoTel and have the same opportunity. But I'm here for the long term. I have confidence in where the company is going, where it is leading to.

Hiring and Training

Because of the rapid growth of the company, the primary job of the personnel office had been staffing. (Senior-level staffing managers reported to the director of human resources, and six field staffing managers reported to the regional sales managers.) In the summer of 1985, employment managers were carrying, on average, a load of 75 to 100 positions before the hiring cutbacks that followed the company's growth adjustment for the second quarter of 1985. Competition for employment was intense: 10 out of 150 applicants who sent resumes got interviews and 1 out of 10 of those interviewed was offered a job. The company's strategy had been to hire experienced personnel.

Until recently, new employees typically did not receive any formal training; there had been neither the time nor the resources. With growth and the development of a separate training center, individual departments had begun to develop more structured training but still essentially adhered to an "on-the-job" approach.

Hiring and training practices varied considerably by area. In sales, for example, although the hiring process was decentralized, all final selections were interviewed and approved by Bennett or Price. Because new sales employees came to the company with experience, training had been limited to introducing the AutoTel product line in a one-week program at the training center. In 1984, a seminar on sales presentations was initiated.

By 1984, it was becoming increasingly difficult to hire the kind of salespeople that AutoTel wanted. Until that time, the company had not given much thought to the idea of in-house development because of the high pace of business growth and the ongoing availability of personnel. However, among other factors, declining unemployment was making it more difficult to find the "AutoTel type." As a result, the company decided in 1984 to hire college graduates as sales assistants and to train them in a twelve- to eighteen-month pilot program. Hired by the regional sales managers, the new marketing assistants were trained at the National Education Center. When the program was evaluated in the spring of 1985, it was clear that it had not met expectations. Among the problems cited were that the training had been too technical and that it had lacked emphasis on the larger picture. In addition, the training responsibilities of the managers to whom the marketing assistants had been assigned were not made clear. As a result, the managers had paid more attention to their quarterly quota pressures than to training the assistants. In the view of one manager: "We got away from the team concept; key players were not part of the development of the program and never bought into it. As a result, it's not clear whether the program will be continued."

In the technical-support department, job training during the early years had been limited to reading product manuals, attending customer classes, and, after a week, "learning under fire." Though the department had hired only experienced personnel, there was still much training required.

By 1985, technical support began experiencing the same recruiting difficulties as sales. Finding people with the necessary background and the right personality for the support jobs was becoming increasingly difficult. The new manager of employee relations, Rachel Pacquette, had spent the winter and spring initiating a college recruiting program for the support area. Pacquette and a number of support personnel visited thirty college campuses, interviewing computer science and engineering students. They interviewed 400 students, made 20 offers, and hired 12.

While the recruiting process was taking place, the format of the "farm team" program, as it came to be called, was slowly evolving. Conceived as a six-month introduction to all AutoTel products, the program was pruned by senior management to a five-week format. The training program was essentially moved from the classroom into the line organization. Formal training would still take place after the five-week initiation, but the new people would be working and contributing. The program included a review of AutoTel products, work on a project, and a mentor relationship. Among various support employees, there was enthusiastic support for the "home-grown concept" and confidence that the program would be successful.

Performance Evaluation

In 1983, the performance review process at AutoTel was formalized. The procedure was for new employees to be reviewed after three, six, and twelve months, with no raise attached to the first two reviews. After the first year, reviews were to be conducted on an annual basis, or more frequently at the request of either employee or manager. The review process included both a numerical and a written rating. Both the manager and the employee filled out separate forms that were used during the review meeting (figures 5–1 and 5–2).

Experience with the review process varied. One employee in development could remember having only one review in the past four years and had to remind his manager about raises when they were due. Another employee in documentation had had four reviews in the past year and a half and observed: "In development, the managers have technical backgrounds and lack supervisory skills. In documentation, the staff generally have a liberal arts background and seem to be people-oriented. Our reviews are more structured and we have more meetings."

Compensation

The comments of one employee reflected Brad Bennett's philosophy of rewarding performance: "AutoTel has the means and the willingness to reward for good performance. You are treated well and people know this." The compensation system, in combination with the practice of internal promotion, mobility, and the high degree of autonomy and free reign typical of a high growth, entrepre-

---

Employee Name _____ Date _____

Position _____ Dept. _____

Please complete the form before you meet with your supervisor to discuss your performance evaluation. Feel free to add additional comments.

1. What do you like most about your job? What do you like least?

2. What are your greatest strengths? Specifically, how have they contributed to your job performance?

3. In which areas do you think you need improvement? In what ways do you plan to accomplish this?

4. Are you looking to assume greater responsibility now or in the near future?

5. Do you feel that your progress to date matches your career goals and expectations?

Figure 5–1. Performance Evaluation Form (Completed by Employee)

neurial company, reflected Bennett's overall commitment to his employees. According to Bennett: "Money doesn't buy happiness, but neither does poverty. The key is peer and self-respect." The company's attrition rate was extremely low, and enthusiasm for the company was very high. In the words of one customer-support manager:

> It's like riding the crest of a wave. Why would you want to get off? At AutoTel, many employees come from the customers. With the competition, employees leave to work for the customers. Here, we know that the company cares. It's the small things, particularly those that are financially tangible, that don't get overlooked. It's a great place to work.

Compensation was estimated to be in the top 50 to 80 percent of the high tech industry. (The company had no data to verify this judgment, however, other than that they were getting the people they wanted.) Percentage salary guidelines were based on the annual review, which was established in 1983. There were no formal grade classifications, salary ranges, or job evaluation systems.

Bennett had also developed a separate semiannual bonus program based on profits and sales growth. Bonus checks were distributed in equal dollar amounts to all employees (except the salesforce) who had been with the company more than six months. In mid-1984, the distribution was $750. The program was unusual because AutoTel based its bonus on profits and sales and made equal distributions regardless of rank or salary. The message was clear. According to Bennett: "This is a reference business and we want everybody to be sensitive to the sales objective. Right down to the custodian, everybody is an important member of the AutoTel team."

Employee Name _____ Date _____

Position _____ Dept. _____

Evaluation Period From _____ To _____ Review Type _____

1	2	3	4	5

*General**

Technical ability
Motivation
Judgment
Quantity of work
Communication skills
Follow-up
Reliability
Relationship with co-workers

Management/Leadership

Delegation of work
Development of staff
Ability to lead and supervise staff

1. Describe the employee's greatest strengths and accomplishments.

2. What are the specific areas in which the employee shows particular growth potential?

3. What additional experience or education do you recommend to improve the employee's development?

4. What areas of the employee's performance do you feel should be improved?

5. What will you do to ensure that performance is improved?

6. Do you believe the present position matches the employee's strengths and career goals?

7. Using the 1–5 scale, rate the overall performance of the employee.

*Scale: 1 = Exceptional
      2 = Good
      3 = Average
      4 = Fair
      5 = Poor

Figure 5–2. Performance Evaluation Form (Completed by Manager)

The fiscal year-end bonus was tied to performance. It was distributed at the discretion of managers, with no set guidelines, but was related to responsibility and salary. However, a $35,000 per year technician could do something outstanding and get a bonus equivalent to a senior manager. A large percentage of eligible employees received bonuses every year. According to one development manager:

Finance divides the money between the areas, such as development or customer support, and loosely oversees equitable distribution to individuals. It can be a good chunk of money. Despite my words to the contrary, it seems kind of expected. Right now, with the company consistently going well, the bonuses I've awarded have consistently been the same or more than the previous year, and the connection to performance might be lost. If revenues drop, this might be a bubble that will burst.

The company's incentive stock option plan offered grants at the discretion of the board of directors. A minimum grant might be options on 500 shares of company stock. Options were granted to "key employees" throughout the company, not just to senior management. A relatively high percentage of eligible employees had received AutoTel stock as a result. Because stock options were vested over a six- to ten-year period, with a 20 percent annual exercise option, valued employees were "locked in." One employee in development admitted that "on more than one occasion, my stock options have kept me here."

Benefits

In 1981, all full-time employees who had been with the company one year became eligible retroactively for a profit-sharing retirement plan. Company contributions to a trust were allocated in proportion to each participant's compensation. Participants could elect to receive an immediate cash distribution of up to 33 percent of their allocation. Other benefits included a 401(K) plan, PAYSOP, life insurance, medical and dental insurance, and 100 percent tuition assistance (figure 5–3).

The Future

Shortly before the end of the second quarter of 1985, AutoTel made a surprise announcement. Earnings for the quarter would not match the adjusted goal. Instead, the company anticipated a drop of about 35 percent, ending the string of thirty-two consecutive quarters of increasing earnings. It expected to report a net income of $4 million for the first quarter, down significantly from the $7 million that had been forecasted. In addition, the "sacred" operating profit margin figure of 20 percent would drop to 13 or 14 percent.

The day after this announcement, AutoTel stock dropped 25 percent. AutoTel's president stated to the press: "The string of thirty-two consecutive quarters of increasing earnings had to come to an end sometime. In retrospect, it only illustrates what a remarkable accomplishment it was."

When official figures were released later in the month, the company reported a 27 percent drop; but the price of its stock remained unchanged. Price remarked: "It's not what we had hoped for at the beginning of the year, but, given

AutoTel Inc. provides medical, dental, life, and disability insurance coverage for employees and dependents as indicated in the following summary:

Medical Insurance
Choice of medical insurance coverage through Aetna Insurance Company or one of the two available health maintenance organizations (HMOs). Covers employees and dependents.

- Aetna Insurance Company   Fully paid by the company
- Silicon Valley Health Plan   Small contribution required by employee
- Bay Area Health Group   Fully paid by the company

Dental Insurance
Covers employees and dependents. Orthodontia coverage for dependent children only, up to age nineteen. Provided through Aetna Insurance Company regardless of whether employee enrolls in that plan or joins an HMO. Fully paid by the company.

Life Insurance
Equal to two times annual base salary to a maximum of $500,000 of coverage. Covers employees only. Fully paid by the company.

Accidental Death and Dismemberment Insurance
Equal to two times base salary to a maximum of $500,000 of coverage. Covers employees only. Fully paid by the company.

Short-term Disability Insurance
Provides 100 percent of monthly salary for ninety continuous days of disability. Covers employees only. Fully paid by the company.

Long-term Disability Insurance
Provides 50 percent of monthly salary beginning the ninety-first day of continuous disability. Covers employees only. Fully paid by the company.

The following benefits are also provided to full-time employees:

Holidays
- Eleven per year

Vacation
- Two weeks first year of employment
- Three weeks beginning second year of employment
- Four weeks beginning tenth year of employment

Tuition Reimbursement
- One hundred percent reimbursement for job related courses if passing grade is achieved.

Bonus Program
- Discretionary based on performance
- Requires six months' employment to be eligible; available to all employees except those paid by commission.

Profit-Sharing Retirement Plan
- Provision for voluntary contributions up to 10 percent of salary
- Fully vested after seven years
- No minimum age requirement
- One year employment required

Figure 5–3. AutoTel Inc. Summary of Company Benefits

the economy and its impact on the high tech industry, a 13 percent operating profit is not too bad." An article in a local newspaper about the end of AutoTel's long string of profit rises stated that AutoTel's president "cheerfully admits that he used to manipulate sales. Business was so good he occasionally would withhold a sale from one quarter's results so he could record it in the next . . . but no amount of juggling could help the numbers last week." By early November, the stock had declined another 25 percent.

Bennett blamed the earnings decline on a slump in equipment spending, more competition, and customers unexpectedly delaying purchases. Though he would not predict results for the second quarter, he did say he considered the first-quarter results an "aberration" and was optimistic because the future prospects looked excellent. However, according to an investment firm, the company's conservative accounting policies cushioned the decline in revenues. Because the company required that buyers pay annual license fees for software and required leasing fees from customers leasing hardware, much of its revenue was built in and was like an annuity, according to the investment firm. The implication was that little new business was being done.

It was a new feeling for most AutoTel employees as, for the first time, they anticipated a diminished bonus check as well as other changes the company might make. Questions arose about the management of the company. The overriding question was how to preserve what had been key to making the company successful. What should be retained? What needed to be changed? Aspects of the company operations that might be affected included the centralized management structure, hiring strategies, and the absence of formal procedures. How these issues would be resolved, and what role the human resources department would play, were questions that would continue to challenge the company.

6
Compensation Systems in High Technology Companies

George T. Milkovich

Human resource management practices in high technology firms enjoy attention in both the popular and the academic press. According to some writers, innovations in personnel policies and practices match the technological innovations that characterize the industry. If high technology firms represent a new phase of the industrial revolution, as some of the literature suggests, then it is a short step to suggest, further, that the personnel policies of these firms also reflect a new phase in human resource management. As evidence, writers point to fewer organizational levels, less formal personnel policies, profit sharing, stock ownership, egalitarian cultures, employment security, Friday afternoon beer busts, flexible hours, and employee assistance programs.

Nowhere is this innovation said to be more evident than in the approaches high technology firms are taking to compensating employees. In a survey of 105 high tech firms, Balkin and Gomez-Mejia (1985) report that the compensation programs offered by high tech firms differ from those of traditional firms in several ways: (1) they place a greater emphasis on incentives (both individual and unit-based) as a component of the total compensation package; (2) they are more likely to offer stock ownership plans for most if not all employees; (3) they emphasize special incentives and rewards for key contributors; and (4) they provide a wider range of services and perquisites to all employees. According to the compensation directors interviewed in the study, "These pay practices are intended to attract, retain, and motivate employees with skills and knowledge critical to the success of the organization."

To some observers, such high tech compensation practices foretell the future; the rest of industry will follow shortly. Yet the compensation practices of high tech firms may also mirror a not-so-distant past. Parallels exist with the early pay practices of now more mature firms. William Cooper Procter, the founder of Procter and Gamble, inaugurated a bonus plan (based on sales) in

1886 and a stock ownership plan in 1903; guaranteed employment followed, along with such employee benefits as disability, pensions, and other programs. Coverage extended to all employees. A *Reader's Digest* article stated: "Colonel Procter was one of the first industrial captains to accept full responsibility for his employees. All these programs make P&G employees unusually content. Yet some of P&G's executives were suspicious, since it represented a completed revolution" (Tisdale, 1937).

In fact, P&G practices were not unique; several other firms matched and extended such pay practices. In its early years (1913–20), Ford's practices included profit sharing, legal and financial counseling (staffed with six full-time lawyers), medical and dental facilities, and employee athletic facilities and musical programs. The arguments offered by firms in support of such pay programs also have a familiar tone: ensuring a stable, experienced work force; reducing turnover of critical-skill employees; avoiding unionization; and improving employee performance and morale.

Whether current pay practices of high technology firms represent real innovations or are mere reflections is not the issue—although insofar as parallels exist, perhaps valuable lessons can be learned. Experience can be a powerful teacher; but experience may also mislead the unwary who do not differentiate between past and current conditions.

This chapter will examine three aspects of compensation practices for employees with technical degrees in high technology firms. First, basic compensation policies are briefly examined and placed in a high technology context. Next is a discussion of whether compensation policies and practices designed for employees with technical degrees differ from policies of other firms. Finally, compensation policies are examined in a strategic perspective. The chapter also considers whether high tech pay policies are part of the forefront of the next evolutionary phase of human resource management and discusses the implications of recent research for pay policies in high tech firms.

High Technology and Compensation Policies

How do you recognize a high tech firm when you see it? Although there are differences of opinion regarding a definition of a high technology firm, there seems to be general agreement that such firms share certain common characteristics. High technology firms are those that emphasize invention and innovation in their business strategy, deploy a significant percentage of their financial resources to R&D, employ a relatively high percentage of scientists and engineers in their work force, and compete in worldwide, short-life-cycle product markets.

Yet it may be misleading to think of high tech firms as a homogeneous group. Differences exist in terms of market share, earnings, financial viability, product diversification, organizational structure, work force size, and so on. Such diversity provides an opportunity for varying personnel and compensation policies. This commonality/diversity distinction suggests two separate but interesting perspectives. The first is to examine how, if at all, compensation approaches in firms that score high on the high technology dimensions differ from those in firms with lower scores. Another perspective is to examine the variations in compensation practices among the firms classified as high tech. For example, do high tech firms with greater market share or stronger rates of growth have different compensation practices from high tech firms that are experiencing less growth or less success in the market? Do high tech firms with lower turnover of scientific personnel, more stable work forces, shorter times to fill vacancies, or lower unit labor costs have different pay practices from high tech firms with high costs or high turnover? Or do the commonalities in high tech—such as the emphasis on innovation, the percentage of scientists and engineers employed, relative R&D expenditures, and short product life cycle—create common approaches in compensation? And do we know if these approaches have discernible effects on the financial or human resources performance of the firms?

Thus, there are really two basic questions that come out of an examination of what constitutes a high tech firm. First, do compensation practices of high tech firms differ from those of other firms? The second question has three parts: Do high tech firms differ among themselves in their approaches to compensation? If so, what do we know about the factors that may influence these differences? Finally, do these differences have any effects on the human resources and financial performance of these firms?

Differences in Compensation Policies

A task of all policy-level managers is to design and orchestrate compensation policies and practices. However, to be able to determine whether differences exist in their approaches, a basic set of pay policy decisions must be identified to compare across firms. Four policy decisions basic to the design of pay systems are considered: the mix of pay forms, competitive position, internal hierarchies, and performance emphasis. Obviously, there are other administrative policy issues that are relevant to designing pay packages for scientists and engineers, including the degree of centralization in corporate versus divisional pay practices and the extent of employee participation and choice in the design and administration of pay. However, the four basic policy decisions listed here permit comparisons among firms.

Mix of Pay Forms

Pay *mix* refers to the various forms of pay in the total compensation package offered to employees. For scientists and engineers, typical forms include base salary; several types of incentives linked to performance, such as merit raises, bonuses, and stock options; and an array of benefits, services, and perquisites.

Policy options facing managers include three aspects of the mix among pay forms: (1) the number of different forms to offer—from merit pay, incentive stock options (ISO), nonqualified stock options (NQSO), and restricted stock to interest-free loans, company cars, and the like; (2) the relative importance to place on each form—that is, the proportion of total compensation deployed to incentive versus base pay, long versus short term, taxable versus nontaxable, cash versus equity, and so on; and (3) the proportion of the work force to be eligible for each form—that is, special treatment for scientists versus all employees eligible for all forms—beyond the legal requirements.

Competitiveness

Pay *competitiveness* refers to the position of a firm's pay relative to the pay of its competitors. Simply stated, a firm can choose to match, lead, or follow its competitors. However, the policy choice is more complex. The pay practices of product market competitors must be identified, since competitive unit labor costs are important; the practices of labor market competitors must also be included, since high tech firms compete for technical employees. Further complicating the policy choice is the widening array of pay forms offered by firms. Some firms seem to emphasize generally riskier forms of pay with high potential payoffs (incentives, stock options), while others emphasize less risk by offering relatively high base pay. Thus, establishing policy regarding pay competitiveness includes assessing the practices of competitors for customers as well as competitors for technical employees; policy must position the total pay package in the market.

Internal Hierarchies

In response to worldwide competitive pressures and the desire to stimulate productivity by reducing corporate bureaucracies, firms are reducing the number of reporting levels. These leaner corporate structures still require attention to pay relationships among jobs and skill levels—the internal wage distribution of the firm. Policy decisions related to internal hierarchies involve the internal equity among jobs as well as the pay of technical/scientific employees in R&D compared to those with managerial responsibilities. In high tech firms, we would expect to find policies regarding internal pay relationships that reflect the emphasis on R&D and the importance of key scientists and engineers to the firm's success. Translated into practice, we would expect to find dual career paths, with the scientific and technical ladder paid at least on a par with the managerial ladder.

Performance Emphasis

Some argue that for high technology firms, performance compensation is an important ingredient for gaining a competitive advantage. Examples used to support this position include the design engineer who is lured away from a well-paid, secure job to a new venture by a package of pay incentives that includes equity in a new firm with a bright future; the team of computer scientists who are offered substantial bonuses to design a new software package within a tight time frame; and the large, diversified firm that rewards innovation and risk taking by technical employees and managers responsible for getting new high tech products to market. Programs that pay for performance—such as profit sharing based on division profits or bonuses to key technical contributors—may be added to the pay package. The premise underlying these examples is that properly designed incentive plans attract talented and motivated personnel.

The policy choices associated with the design of these performance pay programs typically fall along three dimensions. The first involves the age-old problem of short-term versus long-term tradeoffs. Here the mix of various forms of performance pay ranges from merit increases and cash awards to key performers in recognition of their short-term successes to stock options intended to focus employees on the longer-term success of the firm.

Risk aversion versus risk taking is a second dimension of performance pay systems. Some firms try to use pay to encourage managers and scientists to take risks to develop new products and complete projects ahead of schedule. Such pay systems are based on the premise that higher incentives encourage greater risk taking. Others concentrate on using pay to reinforce financial security and reduce the economic risk managers and R&D employees face. However, the meaning of risks and incentives seems to vary among individual employees, and definitions and calculations are not always well defined by either consultants or researchers. A third aspect of a policy decision regarding performance is related to the type of criteria used to evaluate performance. Performance criteria are typically considered quantitative (market share, sales growth, profitability) and qualitative (merit ratings, project completion), based on individual versus group or organization unit and oriented toward entrepreneurial innovation (new product development suggestions) versus manufacturing (gain sharing, cost savings).

These four basic policy areas—the mix of different pay forms, the competitiveness of the total pay package, internal pay hierarchies, and the nature of performance pay—will be used to identify variations in pay systems among firms.

Differences among Firms

A few surveys, mostly by consulting firms, have studied compensation differences among firms. Unfortunately, most of the available data focus on rather narrow, special issues, such as the nature of long-term incentives or base salaries paid. No single survey has reported a comprehensive analysis of the four major pay policy decisions discussed in the preceding section.

Table 6–1 merges the findings from various sources, placing a firm's relative emphasis on high technology in the context of the four pay decisions. According to the various surveys, the mix of pay forms varies between firms with extensive high technology and more traditional firms with limited high technology. High tech firms tend to emphasize incentives and to extend the coverage of incentives, particularly short-term bonuses and stock ownership, to lower levels in the organization. Traditional firms tend to emphasize base salaries and traditional merit increases and to limit the incentive and stock option eligibility of employees to upper levels in the organization and key technical employees. The Hay Group (1985) recently reported that high tech firms are moving bonuses lower into the organization: "In smaller companies, an astounding 47 percent of professional-technical-managerial employees of firms in the under-100 million dollar sales category were eligible for bonuses." Experience suggests that the usual eligibility figure for larger, more mature high tech firms runs between 5 and 10 percent of the managerial-professional-technical work force.

Table 6–1
Compensation Practices across Firms for Technical and Managerial Employees

	Degree of High Technology	
Pay Policy Decisions	*Extensive*	*Limited (Traditional Firms)*
Mix of pay forms	Incentives and base salary emphasized All technical employees eligible Wide range of "perks" and services available	Base salary and merit pay emphasized Limited eligibility by organization level Wide range of "perks" and services available
Competitive position	Most tend to meet competition, some lead	Most tend to meet competition, some lead
Internal hierarchies	Widespread use of formal job evaluation Dual career paths for technical employees	Widespread use of formal job evaluation Limited use of dual career paths
Performance emphasis	Widespread use of bonuses for all technical employees Moderate use of profit sharing Widespread use of stock ownership plans Moderate use of long-term stock options	Limited use of bonuses for technical employees Limited use of profit sharing Limited use of stock ownership plans Limited use of long-term stock options

Sources: Hay Group (1985), Balkin and Gomez-Mejia (1985), Peat Marwick (1985).

Little variation has been reported in the competitive policies of firms. Balkin and Gomez-Mejia (1985) compared the pay systems of thirty-three high tech firms with those of seventy-two "traditional" firms. They classified high tech firms as those that reported their R&D budgets as 5 percent or more of sales. Companies with lower expenditures on R&D were grouped together as traditional firms. In that study, 40 percent of the firms in both the high tech and the traditional groups reported pay policies in which they matched the competition. Approximately 20 percent in each group reported that they set lead positions, and traditional firms were more likely to say that they lagged or followed their competitors' pay. This information is based on the reports of compensation directors; the mechanics used to translate a policy into practice may vary by specific firms. For example, about two-thirds of the firms, both high tech and traditional, reported that they matched their range midpoint with the market's fiftieth percentile. As noted earlier, establishing a competitive position depends on which competitors are included in a firm's calculations of its market's fiftieth percentile. Thus, a lead policy in a market with a large number of firms may be equivalent to a meet-the-competition policy in a market survey that is limited to firms in the same product markets.

The internal hierarchical pay relationships for employees in technical and managerial positions also seem important to all firms. This is attested to by the reportedly widespread use of formal job evaluation plans among technical and managerial positions. Balkin and Gomez-Mejia (1985) did find a major difference in high technology firms: they make more frequent use of dual career paths, which offer technical employees pay comparable to that of managers.

Finally, the surveys by Peat Marwick (1985), the Hay Group (1985), and Balkin and Gomez-Mejia (1986) agreed that bonuses and other forms of performance plans are increasingly part of high tech firms' pay packages. Bonuses and stock ownership tend to be the most widely used, followed by profit sharing and long-term stock options. All surveys consistently reported that eligibility of technical and managerial employees for all forms of incentives is considerably higher in high tech firms.

It appears that high tech firms' pay policies and practices differ in certain important respects from those of firms classified as traditional. High tech firms tend to (1) place greater emphasis on all forms of incentives in their total compensation packages, (2) offer these incentives to a greater proportion of their technical/managerial work force, and (3) design dual career ladders in which R&D work is paid comparably with managerial assignments. These policies seem consistent with the key characteristics used to define high technology firms.

However, the evidence also suggests similarities between high tech and traditional firms as well as significant variations within the high tech group. For example, the majority of firms in both groups set their pay policy to meet their competitors' pay, and both groups make widespread use of formal job evaluation. Significant differences in pay policies within the high tech group were reported by Balkin and Gomez-Mejia (1986). Their data show that over 45 percent of the high tech firms do not use dual career paths and that 20 percent of both the high tech firms and the traditional firms adopt policies that lead rather than meet those of their competitors. Clearly, it is an oversimplification to imply that all high technology firms have similar pay practices. There appears to be as much variation among the pay policies of the high tech firms as among other groups. What is it about these firms, beyond the strategic emphasis on high technology, that influences their pay policy decisions?

Strategic and Environmental Effects on Compensation Systems

At the beginning of this chapter, the pay policies of firms emerging at the turn of the century—such as Procter and Gamble and Ford—were compared to current high tech policies and practices. One inference that could be drawn from this comparison is that history repeats itself. Another possible interpretation is that firms appear to follow a natural life cycle—they are formed, they mature, and they decline. A third explanation is that certain similarities in the nature of these firms and the environments in which they operate may underlie some of the similarities in their pay practices. Accordingly, it may be misleading to assume, as some have, that the compensation practices found in high tech firms are part of the next phase of the industrial revolution and are the wave of the future. That is, confronted with similar pressures, firms in various sectors of the economy—consumer products, manufacturing, and financial services—may respond with similar pay policies.

This section examines the effects of strategic and environmental factors on the design of pay systems, particularly for technical employees in high tech firms.

A currently popular prescription, found in almost every academic article and consultant's report, is for top managers to tailor the pay system to support the firm's strategy, objectives, culture, and so on. The reasons offered seem persuasive. They are based on contingency notions; that is, differences in a firm's strategies, objectives, and culture should be supported by corresponding differences in compensation policies. The underlying premise is that the greater the congruency, or "fit," between the organizational conditions and the compensation system, the more effective the organization. Further, it is argued that different pay system designs should be aligned with changes in strategic conditions. Speculation on what top managers should do can be entertaining, particularly when advisors do not have to live with the results of their recommendations. Consequently, there is no shortage of definitions of strategy or strategic typologies for classifying firms.

In the compensation field, the strategic notion that has received the most attention is organizational life cycles. Based on biological growth curves, the basic premise underlying organizational life cycles is that organizations emerge, grow, mature, and decline—an evolutionary process all organizations are presumed to follow. This appears to be the most commonly used typology in the compensation field.

Specific characteristics are ascribed to each stage of the life cycle, and certain corresponding compensation policies are aligned with these characteristics. For example, in the start-up stage, firms are described as having a single product, usually developed by a founder/entrepreneur. Cash flow is a problem, earnings and revenues are low, and the human resource focus is external—based on obtaining key contributors and encouraging innovation. The pay policies commonly tailored to a start-up stage include a pay mix that emphasizes long-term incentives, to reinforce behaviors consistent with increasing the firm's value; low base salaries; and low benefits, to conserve cash. Contrast this with a firm in the mature stage: its product line is now diversified, revenues and earnings are strong, and the human resource focus is internal—based on consistency of programs within the organization and the need to control cost and enhance productivity. The pay system typically aligned with this stage includes greater emphasis on base salaries, short-term incentives, and competitive benefits.

Despite the appeal of designing differing pay systems to match different stages, there is little reliable information, beyond common sense, on which to base these decisions. For example, there is no reliable method for determining which stage in the life cycle a firm is in or for identifying the breakpoints between stages. This is important if top managers are somehow to tune pay policy decisions to match shifts with stages.

However, our studies at Cornell of the relationship between firms' business strategies and their human resources strategies raise questions about the causality of the business strategy–pay system design relationships. The common view is that business strategies should influence the design of pay systems; but on the basis of some of our studies of firms' planning practices, it can also be argued that historical pay design decisions affect subsequent business strategy decisions. Consequently, efforts to adapt pay systems to shifts in business strategies are hindered by existing pay structures and practices. For example, Apple's recent efforts to balance its emphasis on new technologies with a stronger marketing and operations effort appear to require adaptation of a pay system that rewards technological innovations to one that rewards marketing and operations expertise. Such shifts may result in a loss of key technological contributors and in intraorganizational strife. Thus, Apple's redeployment in its business strategy was affected by a pay system designed to support an earlier strategy. Further, our studies suggest that many high tech firms simply do not develop well-articulated business strategies; the strategies in these firms simply emerge. They are inferred after the fact from the pattern of decisions already made. This is very different from intended strategies that provide well-developed statements of goals and resource allocation guidelines. Under such conditions, designing pay systems

Table 6–2
**Firms' Life Cycle Characteristics and Compensation Policies:
Comparison of Recommendations**

Firm Characteristics	*Start-up*	*Growth*	*Mature*	*Decline*
Product	Single	Simple line	Diversified	Diversified
Revenues	Small	Medium	Medium to large	Medium to large
Earnings	Low	Moderate	Strong	Moderate
Human resources focus	Attract/retain key contributors; external focus	Attract/retain entire work force; external focus	Cost containment; internal focus	Cutbacks and cost reduction; internal focus
Hay Associates Recommendations				
Management style	Entrepreneur, strong leader	Entrepreneur, business manager	Sophisticated manager	Administrator
Performance measurement	Informal, qualitative	Broad goals, measured results	Specific goals, quantitative control	Quantitative, balance sheet
Rewards	High base salary, with discretionary bonus	High level, job related, incentives for results	Average level, job related, incentives for achievement	Average level, incentives for costs
Arthur D. Little Recommendations				
Management style	Entrepreneur, participative	Market manager, leader	Administrator, guide	Opportunist, evokes loyalty
Performance measurement	Informal, qualitative	Qualitative and quantitative	Formal, quantitative, production	Quantitative, balance-sheet
Rewards	High variable, low fixed	Balance of variable and fixed, group and individual [compensation]	Low variable, high fixed, group	Fixed only

Sources: Hay Associates recommendations are adapted from Galbraith and Nathanson (1978). Arthur D. Little recommendations are from Wright (1978).

consistent with business strategies that emerge appears to be an ad hoc or adaptive process.

Perhaps the most important problem faced when attempting to match pay systems with business strategy is that little is known regarding the "correct" set of policies associated with each strategy. As an illustration, table 6–2 lists two different sets of pay policies recommended for each stage by two well-respected sources. One is based on Hay Associates work in executive compensation (Galbraith and Nathanson, 1978); the other is from an Arthur D. Little report on compensation strategies (Wright, 1978). Although comparisons must be made with care because of possible differences in terms, there are some intriguing similarities and differences. For example, a major difference in the two recommendations is in the role of rewards. Hay Associates prescribes greater use of incentives. The point is that no reliable basis exists for advising which, if either, policy option will pay off in terms of desired employee behaviors and firm performance. These differences illustrate the possibility that more than one set of pay policies is correct for any strategic condition.

What, exactly, is known about the relationship between business strategies and compensation policies? It is clear from recent studies that pay policy decisions for managers and technical employees vary systematically across firms with different business strategies. This result has been reported in several studies using very different samples of firms. Broderick (1985), in a study of 199 firms drawn from a wide array of manufacturing firms, found that pay policies for middle managers varied across strategic types. The policies she studied included those discussed in this chapter. Kerr (1985), in a study of twenty industrial firms, found that it is the process of diversification that exerts the greatest effect on the design of pay systems. He found that as product diversification increases (firms move from start-up to maturity in a life cycle context), there is an increased tendency to use quantitative financial criteria for incentive payments. Further, as firms become increasingly diverse, incentive payments tend to be based more in terms of divisional or operating unit results than in terms of overall corporate results.

Finally, Balkin and Gomez-Mejia (1985) also studied the relationship of strategy and other organizational characteristics to pay policies in their sample of thirty-three high tech firms and seventy-two traditional firms. They reported variations between growth-stage versus mature firms. Their overall results confirm the general point that pay policies vary by strategic stages across high tech as well as other firms. Their analysis also suggests that high tech firms in the growth stage are more likely than other growth-stage firms to have a pay mix that places greater emphasis on incentives for technical employees. Further, they report that smaller high tech firms (sales less than $100 million) in the growth stage are the most likely to emphasize incentives for their technical/managerial employees.

It may be productive to look beyond the abstract concept of strategic stage to more specific organizational and environmental factors that seem to be related to pay policy decisions. From this perspective, the firm's life cycle may not be as decisive an influence on the design of pay systems as its financial conditions, including the size of its revenues, earnings, and capitalization and the nature of its products (single versus a diversified product line). Other important influences may include the personal values of key people in the organization, including the founders, and the legal climate, particularly revisions in tax laws. The use of various stock option plans appears to be more directly related to assumed advantages under tax regulations than to any demonstrated relationship to business strategy and managerial performance. In other words, the design of pay systems is a response to external and organizational pressures rather than part of some naturally or continuously evolving process connected with the evolution of firms.

Concluding Observations

The pressures of both organizational and environmental factors shape the design of pay systems in high technology firms. Surveys have consistently shown that, as a group, high tech firms appear to emphasize incentives as a significant part of the mix of their pay packages and to extend them lower into the organization than other firms. However, closer inspection of the data suggests that compensation systems differ considerably within high tech industry. A principal explanation suggested in the compensation literature is that these variations are related to differences in the life stages and business strategies of various high tech firms. Recent research studies support the view that variations in pay systems are related to differences in business strategies. However, it appears that the strategic typologies commonly employed in the compensation field, such as organizational life cycles, are too abstract and may mask the underlying factors, which include the nature of the product and labor markets in which the firm competes, the specific knowledge and skills crucial to the firm's success, the firm's financial flexibility, the legal regulatory climate (particularly changes in tax regulations), and the managerial philosophies of key executives. It is even possible that existing pay systems, designed in response to some discarded business strategy or in response to some now historic pressures, may inhibit a firm's ability to adapt to changing conditions. What is required are pay policies and practices that are loosely coupled with business strategies, thereby permitting sufficient flexibility for managers to adapt to changing conditions.

7

Key Human Resource Issues for Management in High Tech Firms

Robert C. Miljus
Rebecca L. Smith

Today's interdependent global economy requires that U. S. high technology companies, like organizations in other industries, must deal with the reality of continual change. Global competition, America's recent sluggish economic performance, and the declining market share for U.S. products provide the impetus for change in U.S. management processes. International competitors that once relied on low prices to penetrate markets now produce products that are perceived by corporate and household consumers to be superior in service, reliability, and quality (Abernathy, Clark, and Kantrow, 1983; Koch, 1983). The search for effective alternative managerial strategies and practices is also motivated by dramatic shifts in the nature and structure of the U.S. labor force and American workers' desire for more influence in career and work place decisions (Kerr and Rosow, 1979; Levitan and Werneke, 1984).

To remain competitive in a constantly changing global marketplace, U.S. business managers are beginning to recognize the need to identify and implement new human resource strategies. Manufacturing and service organizations in high technology confront the challenge of global competition and technological change on a daily basis.

The pace of technological change can be overwhelming. An engineer becomes technically obsolete less than three years after completing a baccalaureate engineering program. Product life cycles in high tech are shrinking; new products often have to proceed through the design and implementation phases in less than a year if they are to reach the market in time to be economically feasible for the high tech corporation. Finally, project teams in many high technology companies remain intact for less than two years. These factors demand that teams of technical specialists and knowledge workers are effectively orchestrated by managers who (1) possess the skills to manage the business aspects of new product development; (2) are rewarded for collaborating with other teams throughout the organization; and (3) have acquired the knowledge to manage technological change.

It is the nature of high technology businesses to change. High tech organizations themselves create the processes and products that change the industry. What exists in high technology that might not necessarily exist in other industries is the effect of this **internal** force of change; change in high technology originates from the firms' human resources, the "knowledge workers" or "high-talent" personnel. This internal force of change differentiates high technology industry from other industries. Constant, daily change has become the reality of life within high tech organizations. Human resource professionals can best contribute to the firm's success by acknowledging these aspects of change and by working with line management to identify and articulate an organizational perspective within which the knowledge workers can approach their daily tasks.

In what ways are human resource specialists called upon to facilitate line managers' identification, development, and implementation of strategies that will enable U.S. manufacturers to handle competitive and environmental pressures effectively? How will human resource specialists' skills and contributions ultimately improve the organization's performance and enhance the high tech company's ability to compete successfully in the global marketplace? The effective acquisition, compensation, development, and utilization of human resources is critical to the organization's survival. In high technology firms—driven by innovation, science, and research—it is the acquisition, development, and retention of skilled human resources (particularly high-talent personnel or knowledge workers such as hardware and firmware engineers, computer scientists, statisticians, and material and mechanical engineers) that are paramount to the organization's success.

To assess their current role, we interviewed twenty-four human resource managers throughout the United States; seventeen work in high technology enterprises.[1] The interviews were exploratory and open-ended, averaging over an hour. Most interviews were conducted in person; however, several were conducted on the telephone because of geographic limitations.

The principal focus of the interviews was the human resource managers' perception of (1) relevant external environmental forces and (2) key personnel issues confronting their particular firms. The list of human resource managers is not random; hence, caution is needed in generalizing from the results. Time constraints, economic adversity in their firms, and employer policy prohibitions against publicly sharing information were among the main reasons given by several managers who declined to be interviewed. Overall, the responses of those who participated are quite consistent with our own employment and consulting experience and the literature reviewed for the study reported here.[2]

Throughout the interviews, an array of current human resource activities were referenced by human resource managers: equal employment opportunity/affirmative action issues; wage and benefit costs; human resource planning issues; and performance evaluation processes and threats of increased litigation over employee rights. Three high-priority issues were emphasized consistently by the majority of interviewees:

1. Recruitment and staffing—the critical need to attract and properly place high-talent personnel, or knowledge workers, throughout the organization.

2. Training and development—the requirement for the firm to continually assimilate, upgrade, broaden, and deepen knowledge workers' technical skills.

3. Organization design and development—the need to work closely with line managers to shape and maintain organizational conditions that support innovation, change, and employees' continued high performance.

This chapter discusses these priority issues as they relate to knowledge workers, researchers, scientists, engineers, and business and technical managers in high technology companies. Except for minor variations, the responses from human resource managers in high tech and other firms were quite similar. Their responses are treated as a whole in the analysis.

Recruitment and Staffing

Human resource managers indicated the critical need to successfully recruit and effectively place high-talent personnel. Competition for knowledge workers is severe. The American Electronics Association (1981) estimated a compound increase in demand for hardware and software engineers in high tech firms of over 100 percent during the five-year period 1982–87. Besides high technology companies, academic research centers and manufacturers of automobiles, machine tools and equipment, aerospace equipment, and specialty steel are all competing for a limited supply of recently graduated talent. It has been projected that through the mid-1990s, the great majority of new jobs for technicians, information specialists, numeric-control operators, and computer programmers will be in non-high technology industries (Riche et al., 1983). Companies in these industries are undertaking enormous investments to retool their facilities, retrain their employees, and convert their manufacturing operations to computer-controlled and robotics-based manufacturing systems.

Some of the factors that graduate engineers consider when deciding to accept job offers include the geographic location of the firm; the company's image as a technologically innovative organization; the particular project to which they will initially be assigned; the organization's willingness to recognize the individual's contribution; and the company's compensation policy. In turn, high technology companies are aware of and responsive to the requirements of prospective job seekers. In particular, our interviewees identified the following important considerations as relevant to meeting high recruitment quotas.

Establishment and Improvement of the Firm's College Relations

Several major high technology firms rely on aggressive college recruiting efforts, including equipment and monetary grants to leading engineering schools as well as financial support for faculty research (Fombrun, 1982, 48–49). Corporate-sponsored scholarships and internship/cooperative arrangements for students to work on a company's challenging projects prior to graduation are becoming effective recruiting methods for several firms.

Geographic Location of the Firm

Although high technology employment is located throughout the United States and the search for talent is nationwide, over half of the 13 million jobs in the early 1980s were concentrated in ten states.[3] In addition to the established concentrations of high tech industry on the east and west coasts, there are new high tech areas developing in Albuquerque, Austin, Boulder, Chapel Hill–Durham–Raleigh, Colorado Springs, Dallas, Houston, Phoenix, San Antonio, San Diego, Salt Lake City, Tampa–Clearwater, and Tuscon.

What accounts for the attractiveness of such geographic areas to high technology companies in general and to their human resources in particular? Herbert D. Lechner (1985), vice-president for computer resources and administration at SRI International in Menlo Park, California, identifies the following factors from Lyman Carlson's study of Silicon Valley's takeoff and success: an ample supply of key skills and talent, innovative incentives, and desirable geographic attributes. Close proximity to an academic institution provides a flow of talented scientists, and the industry provides the employment challenges, opportunities, and rewards they seek. Lechner adds that availability of sufficient venture capital and a positive business climate in the local community, which encourages a relatively free exchange of ideas and numerous entrepreneurial opportunities, are also vital.

Geographic locations that afford favorable weather, diverse and readily accessible recreational outlets (for example, water sports, snow skiing), access to the arts (for example, museums, opera, theater, symphony), cultural and ethnic diversity, and exceptional academic opportunities for advanced studies, workshops, and seminars are taken into serious consideration by prospective new hires.

Company Image

Several human resource managers stated that the image of the company is very important in recruiting and retaining good researchers, scientists, and engineers. They want to know if the firm is at the "cutting edge" in terms of its technology and research thrusts. Questions commonly asked are: Is the corporation innovative? Has it achieved solid financial growth?

If the company is perceived as a progressive, fair, and challenging place to work, current employees are likely to attract former school associates and colleagues from other firms. Some companies pay a finder's fee or bonus for leads that result in new hires.

Job Challenge and Recognition of Contribution

In addition to company image and innovation, job challenge and the opportunity to be recognized as a contributor to the organization's success are critical factors in the decision to accept a job and to stay with the company. Al Davis, vice-president of Molecular Computer Corporation has stated:

> Job seekers today almost always ask the same first questions: "What is it I'm going to do?" "How much responsibility can I get first?" "I want to have control over something. I want it to be mine." It's "what are you going to let me accomplish?" They want control over their own destiny, but they still want to be part of the best team. They don't want to be told what to do by the boss and be expected to carry it out robotlike without participating in some of the decision. (Barnett, 1983, 171)

The recruit's perception of the organization's response to these issues is developed further in subsequent sections.

Compensation

As noted earlier and consistently emphasized by the human resource managers we interviewed, innovative compensation packages and inducements are required to attract and retain talented knowledge workers. High starting salaries for new college graduates and significant compensation increases for experienced workers, coupled with generous benefits, are essential. Among the benefits are relocation allowances, sign-on bonuses, front-end paid vacations and settling-in allowances at time of hire, graduate education tuition reimbursement, in-house educational opportunities, assistance with home purchase and sale, parallel promotion ladders leading to challenging and high-status projects, and retirement and health benefits packages.

Long-term equity arrangements have not appeared to be a key inducement for new college graduates to join a particular firm. However, short-term bonuses, incentives, and profit-sharing plans tied to individual and group perform-ance are definitely favored. For example, IBM rewarded forty-eight engineers who helped develop their personal computer disk drive with a total bonus of $1.8 million (Rozen, 1985). The use of longer-term stock ownership arrangements that might generate a high payoff if the company should go public is debatable. One of our respondents stated:

> It's only paper! You have no insurance that it will increase in value. Nor is there any obligation on the part of the majority of owners to go public. So what do you have? A better approach is to pay a competitive wage to everyone and provide good percentage increases tied to performance. Profit-sharing which pays peri-odically is a good idea. It has to be carefully designed and the payout fair.

International Employment Opportunities

Several U.S. high technology firms have located manufacturing and data pro-cessing operations outside the United States in an effort to secure significant financial advantages in their manufacturing operations, develop new markets, and access highly educated and creative engineering talent. McCartney (1983) reports:

> National Semiconductor . . . went to Israel, which has an abundance of engineers, scientists and technicians, and had Israeli engineers develop its 32-bit micro [computer]. Control Data, Intel, Motorola have established research and development centers there. . . . Burroughs . . . and Tati, the biggest industrial firm in India, have an agreement by which Tati will supply Burroughs with software development [engineers]. (p. 117)

The market for high-talent human resources is increasingly becoming a global one.

Although research and development are often retained in the United States, standardized high-volume production is taking place offshore. Buffa (1984) estimates that 85 percent of the 36 million people who enter the worldwide labor force annually are from Third World nations, where high-volume products are manufactured at costs significantly lower than U.S. production costs. More then seventy nations are currently competing to attract industry (Blanchard, 1984; *Business Week*, 1985; McCartney, 1983), including Korea, Taiwan, India, Scotland, Jamaica, Sri Lanka, Singapore, Philippines, Malaysia, and Barbados. Among the major U.S. firms responding are Apple, Digital Equipment, Storage Technology, IBM, GE, Amdahl, Burroughs, Ford, Caterpillar, GM, Alcoa, Dow Prime, Wang Laboratories, NCR, Hughes Electronics, Xerox, and Mattel. As the world grows increasingly interdependent, business firms from advanced industrial nations such as France, Japan, West Germany, and Italy, in turn, expand their presence in the United States. In 1981, direct foreign investment in the United States grew to approximately $90 billion (Horovitz and McClenahen, 1982, 73).

It is estimated that about 400 U.S. firms are now located in Ireland and 650 in Mexico. Both nations provide extensive inducements to high tech corporations. Ireland, for example, offers grants for equipment purchases and plant construction, pays 100 percent of the cost of work force training, and provides a corporate income tax rate that tops at 10 percent, with generous allowances for deductions and depreciation. Equally important, producing in Ireland provides a "local presence" and greater ease of entry into the European Common Market (Amatos, 1985). Typical additional inducements are exemption from import duties on parts and materials and a modern infrastructure consisting of good roads, airports, schools, and telecommunications facilities. Opportunities for U.S. knowledge workers and technical managers to relocate for a period of time in foreign countries both to share their expertise and to learn from others can be offered as an attractive recruiting incentive by multinational high technology corporations.

Training and Development

The process of maintaining, upgrading, and expanding the firm's knowledge workers' technical knowledge and skills is of paramount importance to the human resource managers surveyed. The overall objective is to continually improve the technical skills of the firm's human resources in a deliberate, planned, comprehensive, and timely manner. An effective internal technical education program results in improvements in both the performance of individuals throughout the firm and the company's competitive position in the high technology marketplace.

As indicated earlier, competition for knowledge workers is severe; universities are unable to supply sufficient numbers of engineers to meet the growing demand for knowledge workers who possess state-of-the-art skills. As a result, high technology companies expend vast sums to facilitate acquisition of technical knowledge and skills by their internal labor force. A Carnegie Foundation Special Report (1985) cites a figure of $700 million invested per year for internal technical education programs by AT&T. Similarly, at IBM, the $200 million spent for technical education in 1982 more than tripled by 1984.

Four characteristics of training and development tend to be emphasized in high technology organizations:

1. State-of-the-art, new-technology-oriented, in-house education efforts are taking place. In the high tech area, research tends to take place in industry rather than in academe.

2. There is a severe reduction in the product development life cycle; that is, there is less time between the investigation of the possible application of a new technology to a particular market and the actual introduction of the new product to the marketplace.

3. A variety of technical education methods and techniques are used.

4. Career expectations of knowledge workers—as they are influenced by organizational needs—are increasingly recognized by high technology firms.

State-of-the-Art Knowledge

Science and knowledge are the central forces that drive high technology industry. Ultimately, the end product is the embodiment of that knowledge and information in marketplace goods and services (for example, computer hardware and software, fiber optics, laser technology, biomedicine and genetic improvements, ceramics, and new materials). To be marketable, these products must be manufacturable, reliable, functional, supportable, and user-friendly. They must implement the latest technologies and meet customer expectations.

For the high technology organization's survival, it is imperative that state-of-the-art knowledge, concepts, and skills cut across the entire organization—research and development, marketing, quality control, and manufacturing. The firm's human resources must be continually upgraded to prevent obsolescence. At Hewlett-Packard Company (HP), it is estimated that because of the exponential rate of change in the electronic instrumentation and computation industry, knowledge workers become technically obsolete within three years following graduation from a baccalaureate engineering program. Hence, engineers and technical managers are given opportunities to upgrade their skills in areas such as systems architecture, design methodologies, operating systems, programming languages, and so on.

Product Life Cycles

Given the intense nature of competition, knowledge workers in high technology companies must be capable of researching, designing, manufacturing, releasing, and supporting high-quality products within a very short time. As stated earlier, the life cycles for high tech products have been drastically reduced. To be economically feasible to the firm, the cycle for many products, from R&D to manufacturing release, is now a year or less. At the same time, the proliferation of new products is overwhelming. Thousands of high technology products on the market today did not exist ten years ago. Hewlett-Packard has more than 7,000 electronic measurement and computation products on the market, more than 80 percent of which have been introduced since 1980. To raise the knowledge worker's awareness level of new technologies, market demands, and process improvement methodologies, HP provides education in the following areas: product and process development methodologies, market-focused engineer education, and HP-specific product education. No single engineer has access to all the information or resources needed to determine the best place to begin learning new technologies or developing process skills; hence, a dedicated, organized approach to internal technical education is critical to the high tech firm's ability to adequately maintain and upgrade human resources.

Delivery Systems

As in other industries, high technology companies employ a variety of delivery methods for meeting their training and development needs. These range from individualized computer-based instruction to group instruction in classrooms, research labs, seminars, and workshops. Such instruction is designed and offered both by in-house training staffs, assisted by internal technical managers and knowledge workers, and by professional educators from local universities, trade groups, and professional associations. Many Silicon Valley firms provide graduate engineering education programs that are broadcast to company facilities from Stanford University, MIT, the University of California at Berkeley, Chico State, and the National Technological University.

Career Development

It is increasingly recognized that individualized career development opportunities are essential both to attract and to retain knowledge workers. Such programs are typically integrated with performance appraisals in which employees and their managers jointly examine the employees' strengths, future career options, relevant development needs, and alternative growth methods.

Alternative career growth methods may include individualized programs of study (computer-assisted instruction, special projects, self-paced study) as well as group experiences and formal education programs, with tuition reimbursement support. In addition, multiple career tracks, which include increased financial and intrinsic reward opportunities, are essential in high technology firms. These career tracks may include advancement to senior project engineer or scientist; major project leadership assignments; participation in major start-up projects; and upward managerial promotions and movement into technical marketing responsibilities.

High technology firms that are able to attract, retain, and continually improve their knowledge workers' skills tend to be successful on a long-term basis.

Organization Design and Development

Enabling line managers to shape and maintain organizational conditions that support innovation and high performance is the third trend reported by human resource managers in our study. Human resource professionals can help effect change in high technology industries and can help ensure that workers are prepared to meet the competitive demands of the industry. In principle, they must identify, articulate, and provide means for line managers and employees to understand the nature and requirements of changing work processes within the global marketplace. To do this, the human resource manager needs to understand and be able to (1) describe the organizational systems within which the technical experts and business managers are operating and (2) suggest ways for the organization and its workers to respond to new demands by breaking familiar patterns if necessary.

The organizational conditions that shape a company's ability to achieve its goals and respond to changing environments generally include organizational structure, culture or climate, and task design; worker and managerial skills; and required support systems, such as recruitment and rewards, communications, training and development, and conflict resolution. Each of these conditions may be viewed as a continuum, with "traditional" and "adaptive" systems representing the extremes of each (see table 7–1).[4]

Systems at either end of the continuum are not inherently better or worse than their counterparts. Regardless of the organizational system, the overall business objective remains the same—that is, to engender efficient and effective performance aimed at fulfilling the expectations of the multiple stakeholders in the firm (consumers, investors, employees, managers, vendors, government agencies, and so on). The challenge, as Miles and Snow (1984) point out, is to develop the system and the related organizational conditions that strategically fit each company's external environment and support the realization of its specific business goals.

Traditional Systems

For business firms in relatively stable environments, with mature product lines that may command significant market shares, traditional organizational systems may be quite appropriate. This is especially true where capital-intensive technology prevails and much of the work is routine, repetitive, and highly specialized. In this situation, the role of human resource managers is more maintenance-oriented. Human resource planning is short-term; the goal in compensation is typically to remain abreast with (not to lead) community practices; training for employees and supervisors is narrow and task-oriented; the organizational structure is hierarchical; and the climate emphasizes efficiency, cost control, centralized decision making and minimum risk taking. Interpersonal conflict is

Table 7–1
Organizational Conditions and Systems

Organizational Conditions	*Traditional Systems*	*Adaptive Systems*
Structure (shape and design)	Hierarchical/functional	Flatter; project-oriented; task group; matrix
Job and task design	Specialized; standardized	Enriched (do and think); empowerment
Individual skills and expectations	Limited job scope; repetitive	Cross-training; joint problem solving
Managerial skills and style	Directive; centralized	Participative; team building
Culture/climate	Rational; formal; "one best way"	Egalitarian; entrepreneurial; collegial
Support systems		
Recruitment and reward systems	Extrinsic rewards; job security	Merit; intrinsic rewards; professional recognition
Communications	Downward	Multidirectional
Training and development	Task-specific	Career-oriented
Conflict resolution	Conflict-dysfunctional	Pluralism; conflicts negotiated

Note: These conditions and systems are covered briefly in this paper. For more detail, see Beer (1980), Burns and Stalker (1961), Devanna et al. (1983), Lawrence and Dyer (1983), Odiorne (1984), Steers (1977), and Woodward (1965).

viewed as dysfunctional to predetermined personnel rules and is to be corrected by better employee selection, orientation, and training.

Under traditional system conditions, if the firm's employees are unionized, management's usual strategic response has been one of accommodation to minimize employee discontent and ensure uninterrupted productivity. If major changes should occur in the firm's environment (for example, an abrupt shift in consumer demand, an increase in lower-cost competitors, a shift in government policy, deregulation), a variety of strategic choices confront the firm. Depending on previous relationships, the gravity of the change, and managerial values, management may seek concessions to reduce labor costs; request greater work rule flexibility; threaten to establish nonunion susbsidiaries or move offshore; or entertain joint union–management approaches to problem solving, such as quality of working life and quality circle programs (Kochan, McKersie, and Cappelli, 1984; Landen and Carlson, 1982).

The Need for Innovation

Research has consistently revealed that in highly competitive and dynamic external environments, flexibility and the ability to innovate are key factors in organizational success and survival. In *The Change Masters*, Kanter (1983), concluded:

> The corporations that will succeed and flourish in the times ahead will be those that have mastered the art of change: creating a climate encouraging the introduction of new procedures and new possibilities, encouraging anticipation of and response to external pressures, encouraging, and listening to new ideas from inside the organization. (p. 65)

Similarly, in *In Search of Excellence* Peters and Waterman (1982) found that "America's best run companies" were not only good at consistently developing marketable products and services but were especially adroit at responding to changes of any sort in their environments. Peters (1984) adds that continuous attention to total customer satisfaction on the part of every organizational member, by providing product quality, services, and dependability, is the only effective source of sustainable, long-term competitive advantage.

Koch and his associates at the Federal Reserve Bank of Atlanta (1984) also found a major emphasis on innovation in their study of twenty-two high-performing firms located in the Southeast (including Nissan Motors [USA], Nucor Steel, Delta Airlines, Coca-Cola, Hayes Micro-computer, and Federal Express). Adoption of state-of-the-art production technology was paramount. Such firms work with their suppliers to design more functional equipment and continually seek ways to make their equipment more flexible through computerization. With respect to marketing, they actively pursue strategies that clearly define their comparative advantages versus the competition. And they aggressively develop new market niches or exploit proven or mature markets.

Adaptive Systems

Why are some companies more innovative and effective (high performers) than others? Studies reveal that regardless of industry or organization size, successful innovation can be a carefully planned strategy for survival, made possible by enlightened leadership and a consciously developed supportive environment (Drucker, 1985; Quinn, 1985). Koch and his associates (1983) concluded:

> Those companies on the cutting edge of technological change are also on the cutting edge of behavioral change. Successful high-technology companies are led by enlightened managements. They are disillusioned with traditional corporate structures. They believe there must be a better way to operate a business. The model they provide is one of integrating people with technology to get results. (p. 16)

A spirit of entrepreneurship is pervasive throughout the internal environments of high-performance companies. Every employee, regardless of function (operations, marketing, finance, staff) or organization level (from operative to senior manager), is encouraged to actively seek and strive for opportunities to innovate, to contribute, and to be recognized for his or her contributions. The objective is the timely flow of ideas to enhance product value and to improve both product development and product manufacturing processes. The flow of ideas is not the sole responsibility of individual contributors or corporate staff. However, management must assume responsibility for managing and supporting these development processes. Improvement is achieved incrementally: through interactive learning, with many employees participating in various ways; by sharing ideas, experimenting, and learning from successes as well as failures; by measuring (through the use of statistics); and by continually improving product development and manufacturing processes (Deming, 1982).

The attributes of high-performance environments are summarized by Cooper (1984) in a study conducted by Hay Management Consultants. The study covered approximately 1,200 organizations with an employee population of 2.5 million workers. Based on the evidence, the Hay group was able to differentiate between faster-growth and slower-growth organizations and attributed the difference primarily to two distinct business cultures.

In faster-growth organizations (sales revenue and profits clearly above industry averages), more egalitarian and growth-oriented values prevailed. High performance was recognized with advancement and merit compensation. These organizations expected and received high-quality performance. Employees believed that they were the most important asset and that they were respected by management. Reasonable risk taking and learning were encouraged. As a result, such organizations were more productive and responsible to market opportunities, and this tended to be reflected in their financial performance.

In contrast, in slower-growth organizations, power, control of information, and decision making were centralized among senior managers. Authority and threat of job loss were key levers used by managers to direct human resources. Consequently, resistance to change, adversarial behavior, and lower output frequently occurred.

Similar findings were reported in a 1983 national study of 845 working Americans conducted by the Public Agenda Foundation. Over 70 percent of the respondents strongly endorsed the Protestant work ethic and wanted to do the best job possible, but many said that their work environments were quite negative. William M. Ellinghaus, president of AT&T and a member of the Public Agenda Foundation executive committee, asserted that new managerial systems and new ways of organizing work are needed to reinforce employee effort. The report concluded that traditional work place structures—with clearly defined job descriptions, hierarchical authority systems, and sharp distinctions in pay and status—implicitly convey the message that individual workers are less important to the success of the enterprise than managers are.

The Role of the Human Resource Professional

Human resource professionals in many organizations play key leadership roles in helping line managers shape and maintain the organizational conditions that encourage innovation and high performance. At Motorola, for instance, the personnel staff is credited by their CEO, William Weisz (1985), for "helping to create the transition from a more traditional to a more participative philosophy and approach to management" (p. 33). Weisz adds that this change has been intrinsically and financially rewarding for everyone and has led to improved product quality, reliability, delivery, and reduced cost.

In a similar organizational change at Honeywell, Kanter worked with the director of human resources in their Defense and Marine Systems Group, Avionics Divisions (Kanter and Buck, 1985). They combined extensive data-gathering techniques (questionnaires, interviews, open sensing sessions) with participative planning by line and staff groups. As a result, key personnel activities—such as employee orientation, training, minority affairs, and labor relations—were given greater visibility and were further decentralized to the field. Cross-functional task and participative teams were created to more effectively integrate desired organizational conditions with long-range business planning.

These cases serve as backdrop to a third study involving a major provider of technical research and consulting services to both national and international clients, high technology and otherwise. The human resource manager at this firm, which we call Res/Con Services (RCS), was interviewed in depth by Miljus.[5] Even though RCS is in the initial stages of change, the case is most instructive, since it reveals the compelling need for transformation and the role of the human resource professional.

The RCS work force is composed of high-talent research and scientific personnel backed up by a large number of technical and staff employees. The quality and reliability of its services are considered excellent. In recent years, two major changes have forced RCS management to rethink company strategies:

1. Higher-quality competition—national and international—is entering the market with aggressive price and service tactics. Competitors possess top-notch talent, state-of-the-art research technology, and strong research orientations.

2. Consumers are changing. They still demand quality results on complex projects, but they also want quicker turnaround time at lower cost. At times, they also demand timely assistance on very specialized, narrow projects.

In response to these challenges, the principal issue confronting RCS management is how to effectively transform a highly creative but relatively relaxed, informal, campus-like work community into one that is more responsive to business realities—that is, the need to provide timely and reliable services at competitive costs to clients. Management recognizes that it must preserve reasonable autonomy and project decentralization while concurrently introducing professional business principles. As the human resource manager stated:

> Our traditional industrial base is drying up. . . . We no longer are the sole source provider. Bidding for R&D projects is becoming more competitive. We need to be more sensitive to the survival needs of our clients. We no longer are in the mecca; we no longer can sit and wait for projects to come to us.

RCS management further recognizes that many current managers and project leaders will need to be retrained and develop new skills. No doubt, some employees—managers and researchers alike—will resist work role changes and additional accountability requirements. Therefore, a comprehensive incentive package is being developed to support those who may elect to leave (outplacement, buyouts, early retirement provisions) and to induce many others to remain to carry out ongoing projects and to help shape and implement the transition.

At RCS, a classical organizational change effort is under way (see Beckhard and Harris, 1977; French and Bell, 1984; Beer, 1980). The human resource manager, in a major change agent or leadership role, chairs a transition committee composed of key line and staff managers. The major responsibilities of the committee mandated by the CEO are to design future organizational conditions and to spell out transition stages, schedules, costs, communications methods, and a methodology for monitoring and controlling the transformation. Recommendations are to be made on a timely basis to the RCS executive committee for final consideration. Company leaders visualize a three- to five-year time span for making the required changes, and they are optimistic about the process.

Summary

Based on our interviews with human resource professionals, a careful review of the literature, and our own employment and consulting experiences, three human resource issues stand out in high technology industry: (1) recruitment and staffing; (2) training and development; and (3) shaping and implementing organizational conditions that support change and innovation.

High technology firms are driven by their knowledge workers. It is the knowledge workers who create the products and processes that change the industry; and as technologies change, high tech firms are challenged to continuously maintain, upgrade, and expand knowledge workers' skills. Currently, many human resource professionals in high tech companies anticipate an almost unlimited demand for computer scientists and electrical engineers through the 1980s; however, meeting the increased demand for greater numbers of engineers does not appear to be the central issue for these companies. When examined more closely, the demand appears to be for knowledge workers who consistently demonstrate their comprehension, application, analysis, synthesis, and evaluation of technological changes and/or new scientific discoveries in their daily professional tasks.

A variety of compensation- and organization-based inducements to attract knowledge workers have been discussed. Various individualized and group-centered development strategies to prevent skill and knowledge obsolescence have been presented. Finally, several additional organizational conditions that support innovation have been presented, including adoption of egalitarian work place values, organizational reward for collaborative leadership, open communication, creation of risk-oriented environments, joint problem-solving approaches in task groups and project teams, and the value of implementing continuous process improvement methods.

Human resource professionals must build the same level of credibility that line managers and controllers enjoy within organizations by demonstrating the bottom-line advantages of effective management of their firms' human resources. To do this, human resource professionals must acquire knowledge and skills in the areas of business management, product or service improvement, group process, and the deliberate, planned, collaborative process of organizational change. As his or her participation in company planning grows, the human resource person must not expect a "red carpet" from line management; line managers and engineers may be reluctant to throw out work methods for which they have been rewarded in the past. Ultimately, however, line management must have confidence in the human resource professional's ability to frame the system or organizational context within which work is being performed. The role of the human resource professional in high technology organizations should be to facilitate the passage of the organization's human resources through the processes of continual change.

Notes

1. The organizations participating in this study were:

Accuray Corporation	Bank One Corporation
Actavision Corporation	Battelle Memorial Institute
American Electric Power	Compass Computers, Inc.
Anaconda Advanced	Digital Equipment Corporation
Technology, Inc.	Dresser Industries
Anheuser Busch, Inc.	Gelzer Systems Company
Ashland Oil, Inc.	General Electric (Specialty Materials
AU-Tech, Inc.	Department)

Hyatt Hotels Corporation
NCR Corporation
OCLC (Online Computer
Library Center, Inc.)
RCA, Astro Electronics
Rockwell International

Ross Laboratories
Schuler Corporation
Transamerica Corporation
Transmet Corporation
Westinghouse Corporation

2. Professor Miljus has conducted workshops on organization change and management development with such organizations as American Electric Power, IBM, Lear-Siegler, Rubbermaid, Transamerica, United Telephone, and Westinghouse. He especially acknowledges the contributions to this study of Dr. A. Lad Burgin, Director of Management Training, Transamerica Corporation, San Francisco.

3. Percentages were distributed as follows:

California, 20%
  (1.5 million jobs)
Texas, 14%
New York, 12%
Ohio, 9%
Illinois, 9%

Michigan, 9%
Pennsylvania, 8%
New Jersey, 7%
Massachusetts, 6%
Florida, 5%

4. Adaptive systems are also referred to as "high involvement work organizations" by Lawler (1982) and as "new commitment models" by Walton (1984).

5. Because the firm plans to work through its transformation with minimum fanfare and public attention, anonymity was requested.

Part IV
Industrial Relations
Implications

To a substantial degree, the popular media have portrayed high technology industry as a worker's paradise. Obviously, in many firms and for many categories of workers, this stereotype is very far from reality. The chapters by David Lewin and Everett Kassalow focus on matters that high tech firms and their employers' associations rarely acknowledge as significant issues—unionism, technological obsolescence, and work place disputes. In their chapter, Thomas Kochan and John Chalykoff provide an analytical framework for examining a wide range of human resource strategies, including the often overlooked matters discussed by Lewin and Kassalow.

High technology industries remain a frontier for union organization. Among the newer high tech firms, in such areas as Silicon Valley and Route 128 in Massachusetts, virtually no employees are represented by unions. In some established firms that are engaged in high tech activities, there are pockets of unionized employees, though rarely among engineers and other professionals. The relative absence of unionism does not mean, of course, that many of the objective conditions that made widespread unionism attractive to American workers in the past have changed. On-the-job problems remain—especially those in which such questions as discipline, discharge, layoffs, health, safety, fairness, and due process arise.

The union movement acknowledges the difficulties it has had in organizing effectively in high tech firms. At the same time, unions are displaying evidence of a renewed commitment to demonstrate to employees that the protections that collective bargaining has provided to workers in other industries are also needed and can be obtained by workers in high tech.

Increasingly, nonunion firms are recognizing the need for institutionalized procedures for resolving work place disputes between workers and employers. Such procedures are often touted as effective management strategies to prevent unionization on the grounds that they provide effective on-the-job protection for workers without unions. In chapter 8, David Lewin compares the operation and impact of three such plans. In many respects, the plans are intended to operate much like the procedures in collective bargaining agreements, providing systems of appeal that give employees the right to due process in cases of discipline, layoff, discharge, and the like.

Lewin describes the appeals systems, examines appeals issues, and compares the characteristics of users and nonusers of the procedures. He also checks the subsequent work histories of workers and supervisors involved in appeals in order to compare their performance ratings, work attendance, turnover, and promotion with those of employees who have not been involved. His findings will be of interest to those who are inclined to be optimistic about the utility of such procedures in nonunion settings.

In chapter 9, Everett Kassalow examines a broad spectrum of issues relevant to the unions' stake in high technology development. He reviews the AFL-CIO's general views and positions on the impact of high tech on jobs and the economy. He surveys the impact of technological change on collective bargaining practices, and he considers the problems and prospects for union organizing in high tech firms and industries.

With respect to the AFL-CIO, Kassalow notes its recent decision to encourage its member unions to take into account more fully changes in society and the work place that pertain to such matters as acceptance of worker participation. This may signify a trend away from adversarial collective bargaining, which, in turn, may make unions more attractive to workers in high tech firms.

Technological change and the disruption that often follows have long been a concern of unions. The innovations of high tech firms contribute to worker obsolescence in other industries. Furthermore, employees of high tech firms can be the victims. Kassalow reports on arrangements provided in collective bargaining agreements that protect workers and at the same time provide the flexibility in deployment said to be necessary for adaptation to rapid technological change.

Kassalow also examines the status and prospects of increased union organization in high tech. His analysis embraces the difficulties inherent in trying to organize firms in which the work force is often bifurcated by skill level and education (that is, highly trained scientists and engineers and low-skilled electronics assembly workers). In these circumstances, the community of interest that would support unionization is often difficult to mobilize. In addition, many of these firms are subject to strong competition and short life expectancies. They are often geographically mobile and thus a poor investment for union organizing campaigns. Finally, the author explores the salient issue of what motivates workers to join a union and how the highly touted participative management styles of many high tech companies contribute to the difficulties in organizing such firms.

In chapter 10, Thomas Kochan and John Chalykoff focus on human resource management activities at three levels within the firm in order to begin to develop an analytical and testable theory of human resource management systems. They are interested in what determines these systems, their dynamics over time, and their effects on the goals and interests of employees and management.

The authors argue that systematic patterns cut across the three levels of human resource management activities. They identify two such patterns: one a "new system of human resource management" and the other a more "traditional system." Many of the characteristics of the new system are found in high tech for reasons the authors provide, although they emphasize that the new system is also found in other industries.

Especially important in the development of a system of human resource management are environmental pressures from labor markets, unions, government regulation, and technological change, which affect managerial values and business strategies. As these pressures change, human resource management policies may also need to adjust, however incrementally, to accommodate shifting business strategies. At the same time, adjustments can threaten internal coherence and can create potential for contradictions among policies as well as potential for conflict.

The chapter by Kochan and Chalykoff develops a sense of the dynamism that characterizes high tech's new and innovative human resource management practices, linked as they are at the three levels of the firm, influenced by the firm's movement from one stage of the business life cycle to another, and ever subject to the environmental pressures outside the firm's direct control.

8
Conflict Resolution in the Nonunion High Technology Firm

David Lewin

The lack of union success in organizing employees of so-called high technology firms is well known (Miller, 1984). However, the absence of unionism should not be taken to imply the absence of work place conflict in high technology firms. Indeed, a large body of industrial relations research emphasizes conflict as a basic characteristic of the employment relationship (Kornhauser, Dubin, and Ross, 1954), and unionism represents only one possible institutional form for addressing work place conflict.

Another institutional form that has apparently spread to a majority of large U.S. companies (Berenbeim, 1980) is the nonunion appeal-complaint-grievance system (hereafter referred to as the appeal system). Such systems vary along several dimensions: formality, structure, scope of issues covered, representation, and use of third parties. But how do these systems actually work, who uses them, how effective are they in resolving work place conflict, and what are the consequences for the firms and individuals that use these systems?

Few answers to these questions are available (Scott, 1965; Aram and Salipante, 1981). This chapter examines conflict resolution in the high technology firm through analysis of nonunion appeal systems. Data for the study consist of appeal system files and related personnel records for the 1980–83 period obtained from three large firms: Firm A (a diversified financial services company), Firm B (an aerospace company) and Firm C (a manufacturer of computing equipment). The written records were supplemented by semistructured interviews with managers in each of the firms. To preserve the confidentiality and anonymity of the firms (and interviewees) as well as the integrity of the research process, little additional information about the firms will be provided.[1]

The remainder of the chapter is organized as follows. The first section describes the appeal systems prevailing in the three firms and examines their recent usage. The second section analyzes the characteristics of both users and nonusers of these systems. The next section explores the issues taken up through the appeal systems and relates them to characteristics and levels of the appeal settlement process. The fourth section examines the consequences of appeal system usage and appeal settlement via analysis of postsettlement performance, turnover, and internal labor market data. Finally, some observations are offered on the question of whether nonunion appeal systems serve as substitutes for employee unionization.

Appeal System Characteristics and Usage

Each of the firms included in this study maintains a formal, multistep employee appeal system that, at the first or second step, requires appeals to be put in writing in order to progress through the system. The structure of these appeal systems is summarized in table 8–1. In Firm A, informal discussion of an appeal between the employee and the immediate supervisor constitutes the first step of

Table 8–1
Appeal System Structure

	Firm A *(Financial Services)*	Firm B *(Aerospace)*	Firm C *(Computing Equipment)*
Step 1	Informal discussion with immediate supervisor	Written appeal filed with immediate supervisor	Written appeal filed with personnel officer, who meets separately with employee and immediate supervisor
Step 2	Written appeal filed with functional or departmental manager who gives written response	Written appeal filed with personnel officer; hearing officer appointed to meet with employee's immediate supervisor and employee's representative	Written appeal filed with functional or facility manager who meets separately with employee and immediate supervisor
Step 3	Written appeal filed with facility or business unit manager, who gives written response	Written appeal filed with corporate vice-president for employee relations; VP, another management official, and employee's representative constitute board of inquiry	Written appeal filed with divisional or corporate vice-president for personnel who chairs a management appeals committee
Step 4	Written appeal filed with management appeals committee, which gives written response	Written appeal filed with adjustment board composed of outside arbitrator, management official, and employee's representative; board decisions are binding	Written appeal filed with chief executive officer, who makes final decision
Step 5	Written appeal filed with company vice-president for human resources, who makes final decision	N.A.	N.A.

the appeal-handling procedure. Although this appears not to be the case in Firms B and C, archival and interview data clearly indicate that informal discussion is, de facto, the initial appeal-handling step. In each of the firms, four steps are provided for the filing of written appeals.

Note that the appeal system in Firm A does not incorporate formal hearings or employee representation. However, it does provide for written responses to the employee at each step of the procedure, beginning with step 2, and the company's vice-president of human resources renders final appeal decisions. Firm B incorporates hearings and employee representation in its appeal system and also provides for the arbitration of nonunion employee appeals over certain issues (for example, discharge and demotion). The use of employee representation and arbitration in Firm B may well stem from the fact that grievance arbitration exists for unionized employees of this firm (about 7 percent of the company's work force is unionized), and from the use of arbitration by other firms in the aerospace industry to resolve nonunion disputes. The appeal system in Firm C provides a major role for the personnel function in the resolution of work place conflict. The initial written grievance is filed with a facility or plant-level personnel officer, and the corporate vice-president of personnel heads the management appeals committee at step 3 of the procedure. Firm C is also notable for the use of the chief executive officer as the final arbiter of employee appeals.

How often are these nonunion appeal systems actually used? Data relevant to this question are given in table 8–2, where the appeal rate per 100 employees refers to written appeals and excludes informal discussions. The data show that over the 1980–83 period, the average annual appeal rate per 100 employees was 4.0 in Firm A, 6.6 in Firm B, and 6.2 in Firm C. These rates are well below the rates of grievance filing that have typically been reported in unionized settings (Gandz, 1978b; Slichter, Healy, and Livernash, 1960). Regarding the progression of appeals through the respective appeal systems, larger proportions of appeals are taken to higher steps in Firm B than in either Firm A or Firm C. For the three firms as a whole, about two appeals per 100 employees are taken to step 2 of the appeal procedure, about one appeal per 200 employees is taken to step 3, and about one appeal per 500 employees is taken to the final step (step 4). These appeal system usage rates reflect approximately 1,200 written appeals annually in Firm A, 3,400 in Firm B, and 4,800 in Firm C.[2]

Appeal Systems Users and Nonusers

Who uses these appeal systems and who doesn't? To answer this question, random samples of appeals filed in each of the three firms over the 1980–83 period were selected, and these were matched to personnel records to identify employee characteristics.[3] The resultant data are shown in table 8–3. In addition,

Table 8–2
Appeal System Usage, 1980–83

	Firm A			Firm B			Firm C		
	1980	1983	Annual Average 1980–83[a]	1980	1983	Annual Average 1980–83[a]	1980	1983	Annual Average 1980–83[a]
Appeal rate per 100 employees	4.6	4.2	4.0	6.8	6.7	6.6	5.9	6.5	6.2
Step 2 appeal rate per 100 employees	2.3	1.5	1.6	3.1	2.5	2.8	2.2	2.0	2.0
Step 3 appeal rate per 100 employees	0.6	0.4	0.4	1.6	1.4	1.4	0.7	0.8	0.7
Step 4 appeal rate per 100 employees, to: Human resources vice-president	0.2	0.1	0.1						
Arbitration				0.6	0.4	0.4			
CEO							0.1	0.1	0.1

[a]Figures for 1981 and 1982 are not presented separately but are reflected in the averages.

Table 8–3
Selected Characteristics of Appeal System Users, 1980–83

Employee Characteristic	Firm A	Firm B	Firm C
Age (mean, in years)	32.8	34.3	35.4
Sex (%)			
Male	44	76	68
Female	56	24	32
Race (%)			
Caucasian	86	84	87
Black and other minorities	14	16	13
Education (%)			
<10 years	6	7	5
10 to <12 years	16	18	14
High school graduate	28	29	22
Some college	17	19	20
College graduate (bachelor's degree)	24	14	19
Some graduate school	5	8	11
Master's degree	3	3	7
Ph.D.	1	2	2
Years of work experience with the firm (%)			
<1 year	4	5	3
1–2 years	17	17	19
3–5 years	26	23	24
6–9 years	19	21	18
10–14 years	13	16	18
15–19 years	10	9	8
20–24 years	7	6	6
>25 years	4	3	4
Occupation (%)			
Managerial	4	6	2
Professional	24	23	29
Sales	22	18	23
Clerical	23	11	17
Skilled workers	8	13	10
Semiskilled workers	10	21	13
Low-skilled workers	9	8	6

data on the characteristics of the total (domestic) work force in each of the firms were obtained for the 1980–83 period; these data are shown in table 8–4 (on an annual average basis).

Table 8–4
Selected Work Force Characteristics, Annual Average, 1980–83

Employee Characteristic	Firm A	Firm B	Firm C
Age (mean, in years)	36.4	37.8	38.2
Sex (%)			
Male	39	70	61
Female	61	30	39
Race (%)			
Caucasian	89	87	85
Black and other minorities	11	13	15
Education (%)			
<10 years	4	9	4
10 to <12 years	11	15	11
High school graduate	20	26	20
Some college	22	12	21
College graduate (bachelor's degree)	24	23	27
Some graduate school	16	7	8
Master's degree	4	5	7
Ph.D.	1	3	2
Years of work experience with the firm (%)			
<1 year	5	4	6
1–2 years	19	18	15
3–5 years	24	25	21
6–9 years	23	19	23
10–14 years	15	18	17
15–19 years	8	9	11
20–24 years	4	4	5
>25 years	2	3	2
Occupation (%)			
Managerial	12	9	7
Professional	28	26	24
Sales	16	14	32
Clerical	28	14	17
Skilled workers	5	10	6
Semiskilled workers	6	18	10
Low-skilled workers	5	9	4

For the three firms as a whole—and relative to their total work forces—these data indicate that appeal filers were disproportionately young, male, black, less educated, and employed in blue-collar occupations. About two-thirds of all appeals were filed by employees with less than ten years of work experience (with the present employer), but such employees also represent about two-thirds of the total work force in these firms. Managerial and administrative employees—who comprise 12, 9, and 7 percent of the total work force in Firms A, B, and C, respectively—are substantially underrepresented among appeal filers.

To identify further the relationships between employee characteristics and appeal system usage, correlation and regression analyses were performed. The results of these analyses (which are not presented in detail here) showed that age and managerial/administrative occupational status were significantly negatively associated with appeal filing and that males and blacks (and other minorities) were significantly more likely to file appeals than females or Caucasian employees. The significance, but not the signs, of the regression coefficients on the education variable varied according to the estimating equation used; education was negatively related to appeal filing in all cases. Work experience was not significantly related to appeal filing in any of the regressions, using combined (three-firm) data.

Certain interfirm differences in the characteristics of both the total work force and the appeal filers should be briefly noted. Firm A has a predominantly female work force, whereas Firms B and C have largely male work forces; Firm B's work force is less educated and substantially more concentrated in blue-collar occupations than the work forces of Firms A and C; and Firm C's work force is less experienced and has a substantially larger proportion of sales personnel than the work forces of Firms A or B. Appeal filers are relatively more experienced in Firm A than in Firms B and C, more concentrated in blue-collar occupations in Firm B than in Firms A and C, and more educated in Firm C than in Firms A and B.

Although this quantitative analysis helps to distinguish the characteristics of appeal system users from those of nonusers, it does not help to get at the motivational (attitudinal or perceptual) dimensions of appeal system usage. Why do some employees but not others make use of these systems? To deal with this question, consider the exclusion of informal appeal settlement from the data in table 8–2. Recall that Firm A incorporates such settlement in its appeal system, and that interview data confirm the de facto use of informal appeal settlement in Firms B and C. Further, in Firms A and C, special surveys were conducted during the 1980–83 period to determine the extent of informal appeal settlement. In Firm A, surveys conducted in 1981 and 1983 among the total work force found that 15 to 16 percent of all employees had discussed appeals with their immediate supervisor or manager. Moreover, a 1983 year-end survey of supervisory and managerial personnel in Firm A showed that 34 percent of these personnel believed that they had engaged in one or more informal discussions of appeals with their subordinates during the year. In Firm C, a 1982 survey of employees found that 21 percent reported having had informal discussions of appeals with their immediate supervisor or manager. These admittedly fragmentary data support the view—often expressed in the literature on unionized grievance procedures—that informal appeal "filing" and "settlement" constitutes a significant portion of all appeal activity (Kuhn, 1961; Peach and Livernash, 1974).

Regarding the nonusers or underusers of appeal systems in these firms, two pieces of data are relevant. First, a 1982 survey conducted by Firm C found that among self-reported nonusers of the firm's written appeal system, 17 percent said that they had no employment-related issue warranting an appeal; 14 percent had asked their immediate supervisor or manager for "clarification" of a personnel policy; 21 percent reported having held informal appeal discussions with their immediate supervisor or manager; 18 percent feared a general category of "management reprisal" if a written appeal were filed; 14 percent stated that an appeal wasn't worth filing because "management wouldn't change its mind"; 11 percent said that their chances of promotion would be harmed by filing a written appeal; 7 percent said that company policy excluded from the appeal system certain employment-related issues of particular concern to them; 6 percent indicated that external channels or avenues of appeal (such as court suits) were available to them; and 4 percent stated that the appeal-handling procedure was "more trouble than it was worth." Second, among management respondents in Firm C, for whom a more detailed survey was conducted, 34 percent cited a fear of "management reprisal" and 26 percent cited "reduced chances of promotion" (or the jeopardization of career advancement) as reasons for not filing a written appeal.

Apart from any specific explanation offered by nonunion employees for failing to use their employer's appeal system, the significance of the data summarized here is to support the importance of "unobservable events" in analyzing and assessing nonunion appeal systems. The nonuse of such appeal systems and the reasons given for such nonuse are at least as important for the study of conflict resolution in the high technology firm as the data pertaining to appeal system usage, users, and settlement.

Appeal System Issues and Appeal Settlement

What kinds of issues are taken up in nonunion appeal systems? The relevant data for the three firms included in this study are presented in table 8–5, using a sixfold appeal issue categorization scheme (again, the data are for the 1980–83 period).

In all three firms, pay and work issues constitute the largest proportion of all appeals filed (31, 37, and 29 percent in Firms A, B, and C, respectively). However, in Firms A and C, performance and mobility issues rank a close second, accounting for 29 and 25 percent, respectively, of all appeals filed between 1980 and 1983. In Firm B, discipline issues rank second in terms of frequency of appeal filing, and as many appeals were filed over benefit issues as over promotion and mobility issues. Issues concerning sex and race discrimination accounted for between 5 and 8 percent of all appeals filed in these firms

Table 8–5
Appeal System Issues, 1980–83
(percentages)

Appeal Issue	Firm A	Firm B	Firm C
Pay and work	31	37	29
Pay rate, grade or level	6	7	7
Overtime assignment	3	4	2
Job classification	4	4	6
Work standards	4	5	4
Job assignment	7	3	6
Work hours	5	5	1
Safety and health	2	9	3
Benefits	14	15	18
Vacation	3	3	3
Holidays	1	2	4
Personal leave	5	4	6
Seniority	3	4	2
Medical, life, pension coverage and benefits	2	2	3
Performance and mobility	29	15	25
Performance evaluation	7	2	8
Promotion	8	2	6
Transfer	5	3	5
Layoff	2	6	1
Recall	2	1	1
Training	5	1	4
Discipline	12	18	11
Suspension	4	7	4
Demotion	3	4	3
Discharge	5	7	4
Discrimination	7	5	8
Sex	5	2	6
Race	2	3	2
Supervisory relations	7	10	9
Total	100	100	100

during the period under study,[4] and issues regarding supervisory relations account for an additional 7 to 10 percent.

Concerning specific appeal issues and interfirm differences: In the pay and work category, safety and health issues predominate in Firm B; job assignment issues are high on the list in Firms A and C; and work hour issues are far more frequently raised in Firms A and B than in Firm C. In the performance and mobility category, issues of performance evaluation, promotion, and transfer account for 20 percent of all appeals filed in Firms A and C, but only 7 percent in Firm B, where layoff issues predominated. Issues of suspension and discharge represented 14 percent of all appeals filed in Firm B between 1980 and 1983— about twice the incidence in Firms A and C. A substantial majority of discrimination appeals in Firms A and C involved allegations of sex discrimination, whereas race discrimination appeals predominated in Firm B.

To gain understanding of the determinants of issues raised in the nonunion appeal systems, regression analysis was performed using category of appeal issues as the dependent variable and characteristics of appeal filers as independent variables. The results of this analysis (which are not reproduced here) showed that younger workers filed significantly more disciplinary appeals than middle-aged or older workers; female employees filed significantly more performance and mobility appeals than male employees; blue-collar employees filed significantly more pay and work appeals than white-collar employees; and more-experienced employees filed significantly more benefits appeals than less-experienced employees. Concerning specific appeal issues, which were subsequently regressed on appeal filer characteristics, blue-collar occupational status was significantly positively associated with the filing of appeals over health and safety, demotion, and discharge issues; sales workers were significantly more likely than other workers to file appeals over pay issues; women were significantly more likely than men to file appeals over performance evaluation, training, and especially promotion issues; and more-educated employees were significantly more likely than other employees to file appeals over job assignment, personal leave, and transfer issues. Managerial/administrative occupational position was significantly negatively associated with appeal filing across virtually all appeal categories and specific appeal issues.

How does the level of appeal settlement vary (if at all) by appeal issue? Recall that each of the appeal systems included in this study provides for four steps to process written appeals. Table 8–6 presents data on the level of appeal settlement by category of appeal issue for the 1980-83 period. Consistent with table 8–2, the data in table 8–6 show, first, that about 60 percent of all appeals filed in Firms A and B and about 70 percent of those in Firm C are settled at the first written step of the appeal procedure. Second, the level of appeal settlement clearly varies by appeal issue in all three firms. For example, some 70 percent of the appeals over benefit issues and two-thirds of the appeals over supervisory relations issues are settled at step 1, compared to approximately 53 percent of appeals over discrimination issues and 56 percent of appeals over performance and mobility issues.

Third, performance and mobility, discipline, and discrimination appeals are substantially more likely than other types of appeals to be taken to steps 3 and 4 for resolution. Fourth, the use of higher appeal settlement steps (that is, steps 3 and 4) is far greater in Firm B, totaling 21 percent of all appeal settlements, than

Table 8–6
Level of Appeal Settlement, by Issue, 1980–83
(percentages)

Appeal Issue	Firm A, Step Level				Firm B, Step Level				Firm C, Step Level			
	1	*2*	*3*	*4*	*1*	*2*	*3*	*4*	*1*	*2*	*3*	*4*
Pay and work	59	30	9	2	58	24	13	5	73	16	10	1
Benefits	72	23	4	1	69	21	8	2	74	18	8	0
Performance and mobility	49	38	10	5	58	20	18	4	63	21	12	4
Discipline	55	30	11	4	49	20	21	10	70	19	9	3
Discrimination	49	31	15	5	55	19	17	9	58	28	9	5
Supervisory relations	71	22	7	0	64	23	9	4	69	19	12	0
Average	60	28	9	3	58	21	15	6	68	20	10	2

in Firms A and C, where only 12 percent of the appeals filed during the 1980–83 period were settled at these steps. Recall that Firm B is the only one of the firms included in this study that provides for the arbitration of certain types of nonunion employee appeals. Regression analysis, using level of settlement as the dependent variable and category of appeal issue and employee characteristics as independent variables, confirmed that performance and mobility, discipline, and discrimination appeals were significantly more likely than appeals over other issues to be taken to the final steps of the appeal procedures in all three firms. Women and white-collar employees were significantly more likely than men and blue-collar employees to pursue appeals over performance and mobility and discrimination issues through the later stages of the appeals process.

Some additional comments are warranted concerning final-step appeal resolution. In Firm A, the vice-president for human resources was called upon to decide about 3 percent of all appeals filed between 1980 and 1983. The vast bulk of these issues involved female employees who challenged supervisory and managerial appraisals of their performance, claimed that they had been denied promotion and training opportunities, or alleged that they had been subject to discrimination in these and other areas. Of the fourth-level appeals decided in Firm A over the 1980–83 period, 60 percent were decided in favor of employees, compared with about 40 percent decided in favor of employees at all other levels of the appeals system. In Firm B, 15 percent of all nonunion employee appeals were decided by three-member boards of inquiry, and another 6 percent were decided by adjustment boards that included a neutral, third-party arbitrator. The dominant issues at these levels were discipline and discrimination, primarily involving blue-collar and black workers, respectively. Slightly more than half of the appeals brought to these levels *and* involving these issues were decided in favor of employees, compared with an employee "win rate" of 45 percent for all other issues brought to steps 3 and 4 of the appeal system and 40 percent for issues decided at steps 1 and 2 of the appeal system.

In Firm C, where the chief executive officer constitutes the final step of the appeal procedure, some 2 percent of all appeals filed between 1980 and 1983 reached this level. Claims of discrimination, faulty performance evaluation, and inadequate promotions brought by female employees were most likely to reach the final step of the appeal procedure in Firm C. Slightly less likely to reach this level were allegations of unfair transfers and supervisory denial of personal leave made by relatively more educated and experienced employees of Firm C. Of all appeal issues reaching the final settlement level in Firm C between 1980 and 1983, roughly 66 percent were decided in favor of the employee, compared with about 45 percent at other (lower) levels of the appeal procedure.

These data suggest that, when filing written appeals, nonunion employees of Firms A, B, and C are more likely to prevail—that is, to receive a favorable decision—the further such actions are pursued through the firms' appeal systems. At the same time, the probability of an appeal reaching the final stages of the appeal system is low, but this probability varies significantly by specific appeal issue and characteristics of appeal filers.

Consequences of Appeal Filing and Settlement

The substantial literature on grievance procedures for unionized employees and the relatively sketchy literature on appeal systems for nonunion employees share one important characteristic—a single-minded focus on the grievance-appeal settlement process (Chamberlain and Kuhn, forthcoming; Aram and Salipante, 1981). An equally if not more important concern, however, has to do with the consequences of appeal filing and settlement. How (if at all) does appeal activity affect employee and supervisor job performance, turnover, work attendance, and promotion? Do appeal filers and those against whom appeals are filed suffer in terms of subsequent performance and organizational rewards, compared to employees and supervisors/managers who are not officially involved in appeal activity?

To answer these questions, appeal system and personnel records for samples of employees and supervisors of certain facilities and locations of Firms A, B, and C were selected for analysis. All appeals filed by employees during 1982 in these facilities and locations were examined to identify the issues involved, the levels of settlement, and the outcomes of settlement. Then performance, work attendance, turnover, and promotion data for employees and supervisors/managers involved in appeal activity in 1982 were obtained for the 1981–83 period. Next, comparable performance, work attendance, turnover, and promotion data were gathered for samples of employees and supervisors/managers of these facilities and locations who were not involved in appeal activity during 1982 (again, data were collected for the 1981–83 period).[5]

In all three firms, employees filing appeals in 1982 had higher performance ratings and more work absences than nonfilers over the 1981–83 period (table 8–7). Chi-square tests showed that these differences were insignificant in the case of performance ratings but were significant in the case of work attendance. Second, in all three firms, both voluntary and involuntary turnover was higher among appeal filers than among nonfilers in 1983; these differences were statistically significant in all cases (that is, by type of turnover in each firm). Third, employees filing appeals in 1982 were significantly less likely than non-filers to be promoted in the subsequent year; this was especially true in Firm A.

Fourth, performance ratings, work attendance, employee turnover, and promotions for employees filing written appeals vary substantially—in some

Table 8–7
Personnel Activity for Appeal Filers and Nonfilers

	Firm A		Firm B		Firm C	
Year and Personnel Activity	*Appeal Filers*	*Nonfilers*	*Appeal Filers*	*Nonfilers*	*Appeal Filers*	*Nonfilers*
1981						
Performance rating (1 = low, 5 = high)	3.7	3.5	3.7	3.4	3.8	3.5
Work attendance (% of days absent and late)	7.4	5.9	6.2	5.8	6.0	5.6
1982						
Performance rating (1 = low, 5 = high)	3.6	3.4	3.5	3.4	3.6	3.4
Work attendance (% of days absent and late)	6.9	5.7	6.0	5.6	5.8	5.4
1983						
Performance rating (1 = low, 5 = high)	3.4	3.4	3.6	3.3	3.5	3.3
Work attendance (% of days absent and late)	7.3	6.0	6.4	5.7	6.1	5.7
Turnover (% of all employees)						
Voluntary	6.5	4.2	5.6	4.4	2.7	1.6
Involuntary	3.3	2.4	4.0	2.8	2.0	1.2
Promotion (% of all employees)	3.7	6.2	1.9	2.8	4.3	5.9

cases significantly—by the level to which appeals are taken and by the outcome of appeal settlements (table 8–8). Specifically, appeal filers who took their cases beyond step 1 of the appeal procedure had lower performance ratings and higher involuntary turnover rates in the year following appeal filing than appeal filers whose cases were settled at the initial step of the appeal procedure. The turnover differences were statistically significant in all cases (that is, in Firms A, B, and C), whereas the performance rating differences were significant only in Firm A.

Even more significant differences in personnel activity in these high technology firms are evident from examination of the outcomes of appeal decisions. Employees whose appeals were decided in their favor in 1982 had significantly lower performance ratings, work absence rates, voluntary turnover rates, and promotion rates in 1983 than employees whose appeals had been decided in favor of the employer. Employees who "won" their appeals in 1982 also had significantly higher involuntary turnover rates in 1983 than employees who "lost" their appeals in 1982.

Fifth, supervisors and managers who were party to employee appeals in 1982 differed significantly from supervisors and managers who were not party to such appeals in terms of subsequent personnel activity. Specifically, supervisors and managers against whom employee appeals were filed in 1982 had significantly lower performance ratings and promotional rates and significantly higher turnover rates (both voluntary and involuntary) in 1983 than supervisors and managers against whom employee appeals were not filed in 1982 (table 8–9). These differences were slightly greater in Firm A than in Firms B and C.

Taken as a whole, these findings provide partial support for the view that appeal filing in the nonunion high technology firm can have negative performance assessment and career advancement consequences for employees who file appeals and for supervisors and managers who are party to such appeals.[6] Although no doubt unintended, in the sense that a firm does not consciously seek to impose penalties on the portion of its work force that is involved in appeal activity, the data examined here suggest that nonunion employees who file written appeals do so at considerable risk, and the risk spreads to the supervisors and managers of those employees. These findings help to explain both the reported high incidence of informal appeal settlement and the low incidence of appeal system use by managerial and administrative employees in the three firms included in this study.

Appeal Systems and Employee Unionization

Do nonunion appeal systems in high technology firms in effect serve as substitutes for employee unionization? An answer to this question depends, in part, on how much organizational control can be exercised by managers in these firms.

Table 8–8
Personnel Activity in 1983 for Appeal Filers in 1982

Personnel Activity (1983)	Firm A, Appeal Taken to		Firm B, Appeal Taken to		Firm C, Appeal Taken to	
	Step 1	Steps 2,3,4	Step 1	Steps 2,3,4	Step 1	Steps 2,3,4
Performance rating (1 = low, 5 = high)	3.2	3.5	3.7	3.5	3.4	3.6
Work attendance (% of days absent and late)	7.5	6.9	6.6	6.2	6.4	5.9
Turnover (% of all employees) Voluntary	6.6	6.4	5.7	5.5	2.8	2.6
Involuntary	3.0	3.6	3.8	4.3	1.8	2.2
Promotion (% of all employees)	3.9	3.5	1.4	2.1	4.5	4.1

Personnel Activity (1983)	Decision in Favor of		Decision in Favor of		Decision in Favor of	
	Employee	Employer	Employee	Employer	Employee	Employer
Performance rating (1 = low, 5 = high)	3.2	3.6	3.4	3.7	3.3	3.7
Work attendance (% of days absent and late)	6.8	7.6	6.1	6.9	5.8	6.3
Turnover (% of all employees) Voluntary	6.0	6.8	5.2	5.7	2.5	2.8
Involuntary	3.7	2.8	3.4	4.7	1.9	2.1
Promotion (% of all employees)	2.1	4.1	1.6	2.3	4.1	4.6

Table 8–9
Personnel Activity in 1983 for Supervisors and Managers, by 1982 Appeal System Status

Personnel Activity (1983)	Firm A, Supervisors/Managers with		Firm B, Supervisors/Managers with		Firm C, Supervisors/Managers with	
	Appeals Filed Against Them	Appeals Not Filed Against Them	Appeals Filed Against Them	Appeals Not Filed Against Them	Appeals Filed Against Them	Appeals Not Filed Against Them
Performance rating (1 = low, 5 = high)	3.4	4.2	3.6	4.0	3.8	4.2
Turnover (%)						
Voluntary	1.4	1.1	1.7	1.3	1.8	1.6
Involuntary	1.1	0.7	1.3	1.0	1.1	0.9
Promotion (% of all category)	1.0	1.8	1.1	1.7	2.1	2.9

The history and literature of unionism and industrial relations suggest that such managerial control is quite limited.

The scientific management and human relations philosophies and practices of U.S. management developed earlier in the twentieth century provided no place for unionism in their frameworks and were unable to prevent the subsequent unionization of employees (Bendix, 1956). Similarly, personnel departments that developed early in the twentieth century were not (as their creators and advocates had assumed) regarded by employees as neutral protectors of employee interests (Jacoby, 1981). More recently, management philosophies and practices based on concepts of job satisfaction and quality of working life have not been shown, per se, to supplant or be negatively related to employee unionization.

Indeed, arguments have raged for many years over the determinants of unionization. At one time, industrial unionism was judged to be unattainable, but the post-1935 experience of the United States refuted that judgment. Similarly, the unionization of government employees, hospital workers, professional athletes, and college faculty was once regarded as unlikely to occur—even unthinkable—yet these sectors are now among the most highly unionized in the economy (Troy and Sheflin, 1985). Further, numerous industries and types of firms, including putatively high technology firms, that are not organized in the United States are unionized in other countries (Hill, 1982). None of this is meant to mask the three-decade decline of unionization in the United States (Dickens and Leonard, 1985), the reduction in the proportion of union victories in representation elections (Heneman and Sandver, 1983), or the lowered opinion of unions and union leaders among the public at large. Rather, it is meant to support the point that, although it is difficult to predict, unionization has been shown by most industrial relations researchers to be determined by factors that are outside direct managerial control—that is, by exogenous variables.[7]

In contrast, the organizational behavior and conflict management literatures, as well as more behaviorally oriented industrial relations research, strongly suggest that managers exercise considerable control (albeit perhaps indirectly) over employee unionization. For example, the industrial relations executive has recently been described as a "manager of conflict" (Gandz, 1978a), and conflict management theorists identify the integration of individual and group goals within organizations as a key management function (Brown, 1983). Holding other factors constant, the use of employee relations consultants has been shown to be significantly negatively correlated with union victories in representation elections conducted by the National Labor Relations Board (Lawler, 1984; Lawler and West, 1985). Relatedly, several studies have found that employer-initiated delays in holding union representation elections (such as by challenging the composition of the bargaining unit) and the volume of unfair labor practice charges incurred by an employer are significantly negatively associated with union win rates in representation elections (Cooke, 1983). Consistent with these empirical findings is the view that "it is the nonunion sector that is the source of innovation in the personnel area, including discipline" (Jacoby, forthcoming, 61; see also Vollmer and McGillivray, 1960).

What all of this adds up to is a picture of modern employers—perhaps concentrated in high technology firms but not confined to them—that consciously make investments in human resources, employee relations, and industrial relations practices that are intended to ward off, contain, or displace employee unionization and that have achieved such effects. This conclusion seems to be supported by the data from the study reported here, which show a nontrivial use of appeal systems by employees of three high technology firms but also show certain negative consequences of appeal system usage and settlement for employees and their supervisors/managers. It is also supported by a recent Conference Board study (Freedman, 1985) which found a significant positive correlation between large American companies' use of nonunion employee appeal systems and management's desire to remain nonunion (or partially unionized). Finally, the exceptionally high voluntary employee turnover among employees of high technology firms, and the related fact that voluntary turnover ("exit") is the prime alternative to the exercise of "voice" (largely through unionism) in the employment relationship (Hirschman, 1970; Freeman and Medoff, 1984) support the view that managers can indeed exercise a good deal of control over employee unionization via the choice of conflict management and human resource policies and practices.

It is uncertain which of these competing hypotheses—that managers exercise fundamental control over employee unionization and can devise substitutes, such as appeal systems, for it or that the determinants of unionism are basically exogenous and are not subject to managerial control—will ultimately be validated by empirical research. This chapter is intended to help stimulate further research on this question as well as on the processes and outcomes of nonunion appeal systems and other conflict resolution devices in high technology firms.[8]

Notes

1. This condition was agreed to by the researchers as part of an agreement concerning access to and use of the data. Limited additional descriptive information about the firms is available from the author upon written request.

2. Based on the average annual domestic employment of these U.S.-based firms over the 1980–83 period.

3. The randomization process was applied to appeal system files within specific plant, facility, and corporate locations of each of the firms. Selection of plants and facilities was made on the basis of discussions with human resource and industrial relations executives and data availability. Hence, the plants and facilities included constitute a purposive sample.

4. Two of the three firms included in this study also maintain special appeal and

alternative dispute resolution procedures to deal with discrimination cases. However, the first (written) step of the appeal systems discussed here must be used for such discrimination cases, so the data in table 8–5 do not appear to significantly understate the proportion of all appeals represented by discrimination appeals.

5. The specific sample selection procedures and methods of assembling the data follow closely those that were recently used to study the consequences of grievance settlement in unionized settings in four industries and sectors (Lewin and Peterson, forthcoming).

6. For evidence that grievance filing is significantly negatively associated with plant-level productivity in pulp and paper mills, see Ichniowski (1984). For further examination and details of the analytic methods used to study relationships between appeal and grievance activity and employee performance, see Lewin and Peterson (forthcoming).

7. See, for example, Ashenfelter and Pencavel (1969) and Moore and Newman (1975).

8. This is consistent with the recent emphasis on behavioral research in industrial relations (Lewin and Feuille, 1983).

9
The Unions' Stake in High Tech Development

Everett M. Kassalow

The U.S. trade union movement has been caught up in a sea of change in the past decade. Not all of this change is technological, but it is difficult to disentangle the technological from the structural, demographic, or cyclical. From a practical union viewpoint, labor must respond to all of these changes and many others, making fine distinctions impossible. This chapter will concentrate on the problems of unionism in a high tech age.

High tech developments infringe upon unionism in a variety of ways. For convenience, the chapter has been divided into three sections: the first reviews the AFL-CIO's general views and positions on the impact of the high tech revolution on jobs and the economy; the second surveys the impact of recent technological change on collective bargaining practices in the United States; and the third looks at some of the problems and prospects for union organizing in high tech industries.

The AFL-CIO View of the Impact of Technological Change on Jobs and Wages

In its so-called Florida report—the second report to be issued by its Committee on the Evolution of Work—the AFL-CIO (1985) stated: "The United States—indeed every industrialized nation—is undergoing a scientific, technological,

My thanks are due to a handful of national unions and union officials who have helped me in a number of ways. In addition to the national unions that I have acknowledged in notes, I should like to thank the Communications Workers of America and the following AFL-CIO officials: Charles McDonald of the Organizing Department, Mark Roberts and John Zalusky of the Research Department, Dennis Chamot of the Department of Professional Employees, and Richard Prosten of the Industrial Union Department. I need scarcely add that none of these people is responsible for the interpretations and conclusions of this chapter; indeed, they might surely disagree with some of them.

economic revolution every bit as significant as the industrial revolution of the nineteenth century" (p. 6). The AFL-CIO has taken a fairly gloomy view of that revolution and the way in which it is altering jobs and wages—claiming, in its first report, that the new technology "is changing the structure of the U.S. economy, changing the foundation of national power and national security— and also contributing to the nation's labor surplus" (AFL-CIO, 1983, 7). Whatever benefits it brings, advancing technology also brings disproportionate social and economic costs, including the downgrading and displacement of workers, plant shutdowns, joblessness, and more workers working at lower pay.[1] In addition, the AFL-CIO believes that there won't be enough new high tech jobs created to replace those lost in declining industries.

Union Fears of a Two-Tier Work Force

The AFL-CIO believes that with computers and robots taking over more and more functions in the factory and office, the general result could be the creation of a two-tier work force, in which some jobs will be upgraded but many more will be downgraded. As a result, there will be a job structure with relatively few "executives, scientists and engineers, professionals and managers" at the top who will decide "whether the work will be done by people or robots [and] in the United States or overseas." At the bottom of the job structure will be "low paid workers [doing] simple low-skill, dull, routine, high-turnover jobs." Sandwiched between the top and bottom will be fewer and fewer of the traditional well-paid blue-collar "skilled, semi-skilled and craft production and maintenance jobs, which in the past [have offered] opportunity and upward mobility to . . . workers who start in low paid, entry level jobs" (AFL-CIO, 1983, 8). Thus, the AFL-CIO analysis links recent technological change with the steady loss of jobs and technology overseas.

It is not just the elimination of well-paid manufacturing jobs that union leaders fear. There are similar trends in office work. Addressing a conference on office work and new technology, Karen Nussbaum, director of 9 to 5, the National Association of Working Women, argued in 1982 that automation could "lock in" clerical workers to low paying jobs, and leave them without career advancement opportunities (Bureau of National Affairs, 1984c, 6).

Whether this view of an emerging two-tier economy is valid could be the subject of a separate paper; but a few comments are in order. A few years ago, Eli Ginsberg noted that the great bulk of the new jobs established in the 1970s were of a low-level, low-paid service type.[2] Whether this trend will continue to be the reality of the last half of the 1980s remains to be seen. The birth rate bulge of the 1950s and early 1960s has expended itself, which means a smaller entering labor force cohort in the late 1980s. Thus, declining rates of labor force increase offer the possibility of tighter labor markets and higher capital labor ratios, along with rising productivity per worker in the expanding service sectors. This could lead to higher wages, even in the traditionally low-paid jobs—a process that could be heightened by a rising rate of unionism in the private sector. In any event, the AFL-CIO persists in its pessimistic view of these processes, and this viewpoint affects its outlook and positions.

To counter these economic and social trends, the AFL-CIO proposes a broad program of measures to restore full employment, to reduce work hours, and to pursue a more realistic trade policy. This program is consistent enough with past positions the organization has taken, although its stand on trade was different two and three decades ago.

Pessimism, declining numbers, and the inability until now to make a major impact on organizing in the newest high tech sectors of the economy place the AFL-CIO in a different role from the one it played two or three decades ago. In the early 1950s, on the eve of the greatest period of economic growth in recent U.S. history, organized labor was in the forefront of promoting an optimistic view of the economy. Today, while the AFL-CIO continues to struggle for full employment, its weak foothold in high tech and its efforts to fight off shrinkages in auto, steel, rubber, and other sectors leave it with a far different image from that of the 1950s.

The Change in Labor's Outlook

One of the most interesting signs of change and evidence of a new will to make adjustments to a changing society, including changing technology, is a greater and more explicit acceptance by the AFL-CIO of the importance of job satisfaction and quality of work life programs. The 1985 report by the AFL-CIO Committee on the Evolution of Work suggests that "the labor movement should seek to accelerate this development," while cautioning that "some employers have used quality of work life programs as 'union avoidance' measures or as simple 'speed-up' efforts" and insisting that only where such programs are "grounded in collective bargaining" can workers be adequately protected from possible abuses (AFL-CIO, 1985, 19). Even this skeptical and limited acceptance of quality of work life programs is a long step from the earlier hostility and complete rejection of such programs.[3]

In its quest to reach a changing work force, the AFL-CIO is deemphasizing its traditional adversarial approach to labor–management relations. Its latest report on the future of work notes that "many workers, while supporting of the concept of organization, wish to forward their interests in ways other than [in] an adversarial collective bargaining relationship." This same report also comments: "We understand confrontation and conflict are wasteful and that a cooperative approach to solving shared present and future problems is desirable" (AFL-CIO, 1985, 6, 18).

The AFL-CIO's Tom Donahue had described collective bargaining as a "conflict relationship, where equals talk with equals." In that same statement, he added: "We don't seek to be a partner in management . . . we do not want to blur in any way the distinctions between the roles of management and labor in the plant" (AFL-CIO, 1976, 6). Donahue, now secretary-treasurer of the AFL-CIO, has been the federation's leader in the reexamination of traditional attitudes and serves as chairman of the committee that produced the recent reports on the future of work.

The trend away from adversarialism is now evident in the machinery, transportation equipment (including automobiles), and transportation industries, where unions and management have initiated management worker or union representation on company boards of directors. Union–management experiments with employee stock ownership programs are also on the increase. For the most part, these experiments are limited to companies that have been experiencing serious economic difficulty. Although there is no general endorsement of these moves by the AFL-CIO, they do represent an interesting departure from the traditional adversary relationship.

It is not easy for labor to modify its traditional adversary philosophy in an era when employer opposition to unionism, especially in unorganized work sites, is, if anything, greater than ever. Yet in companies where unions have long been established—such as Ford and General Motors (GM), American Telephone and Telegraph (AT&T), and in basic steel—contract language inserted in recent years also attests to a new spirit of cooperation. A clause in the agreement between the Communications Workers of America (CWA) and AT&T is directed specifically at technological change:

> The Company and the Union recognize that technological changes in equipment, organization, or methods of operation have a tendency to affect job security and the *nature* of the work to be performed. The parties, therefore, will attempt to diminish the detrimental effect of such change by creating a joint committee to be known as the Technology Change Committee to oversee problems and recommend solutions of problems in this area. (AT&T–CWA, 1980, 29–30)

Under this arrangement, the company is to notify the union six months in advance of planned major technological changes. Both parties would jointly explore the impact on employees of retirement and severance benefits and transfers and retraining possibilities for affected employees. In addition, the major auto company agreements (with the UAW) in recent years have established joint technological change committees. The GM–UAW agreement recognizes that technological progress and advances in living standards rest on "a cooperative attitude on the part of all parties" in dealing with technological change (GM–UAW, 1982, 430–31). Originally, the major union objective in these national and local committees on technological progress seemed to be to deal with the way new technology might affect the size of the bargaining unit.

According to UAW Ford Department officials, however, in recent years, the committees have provided the basis for wider cooperation between the parties in coping with technological change.

Unionism, Bargaining, and Technological Change

Conventional wisdom in the past has held that in the face of technological change, craft unions were likely to present much greater resistance than industrial unions. Craft unions feared that major technological advances might shift work away from their members, since new skills or materials could be involved and an entire craft could be threatened with extinction.

The same wisdom found that there was less resistance to change among industrial unions, since "technological changes affect such a small fraction" of an industrial union's membership "or affect different parts of the union in different ways." Indeed, "change that hurts some [industrial union] members . . . may help other members." Consequently, ran the argument, industrial unions were "likely to pursue a policy of adjustment" to technological change, rather than resist it (Slichter, Healy, and Livernash, 1960, 345). A combination of long-run structural changes in auto and steel markets has led unions in those industries to pose new demands and devise new programs. They and other industrial unions have had to deal with (1) shutdowns of older, technologically obsolete plants; (2) drastic reductions in employment opportunities of their members in still-functioning plants as a result of new technology; and (3) new forms of work organization (and, of course, changing markets). Some industrial unions have also found that new technology may threaten to shrink the jurisdiction and scope of long-established bargaining units. In responding to these and other aspects of technological change, the unions have often been quite innovative and have broken some new ground—as, for example, in establishing new training programs or in helping to set up new systems of job security. All of these developments have led to an increase in what might be called the "scarcity consciousness" of a number of industrial unions—a kind of consciousness more identified with craft unions in the past.

The newness of the technological issues confronting the unions today can be exaggerated. At a special AFL-CIO conference on technological change a few years ago, the president of the Communications Workers of America observed: "The old problems are still with us." Many of the problems stemming from today's technological changes are "the kinds of things that we were talking about 30 and 40 years ago when we're talking about simple mechanization." And, he added, the same "problems will be with us 20 years from now on a different scale" (see Chamot and Baggett, 1979, 31).

In a somewhat similar vein, the president of the Brotherhood of Railway, Airline and Steamship Clerks remarked: "In the early years rail unions fought *for* technological change." Work was harsh and dangerous, and "anything that would improve things was welcome." But after World War II, "unions in the rail industry started fighting against technological change" or, rather, "fighting against its adverse effects" (Chamot and Baggett, 1979, 32). As is the case with many industrial unions today, it was the threat to employment and members' job security that dictated changes in the unions' posture.

However, it is not only in the industrial, manual worker area that new technology has challenged the bargaining practices of U.S. unions. Technological invasion of white-collar work has also posed new and difficult bargaining issues. Video display terminals (VDTs) and other electronic devices have become commonplace in the office and related areas. Labor unions were almost the first to react to these developments, and they have devised programs and responses to meet some of them. Since it is impossible to trace all of the collective bargaining responses, we have picked out some important examples to illustrate the ways in which bargaining units are trying to come to terms with technological change.

The Bargaining Units

The CWA reports that while computerizing some skilled operations, AT&T not only simplified them to require lesser skills and fewer workers but also tried "to remove part of the job from the bargaining unit as a management control function." The CWA charges that AT&T sought to monopolize knowledge of the new work and refused "to train members of the bargaining unit, thus robbing the unit of both knowledge and work." In this case, the union won "arbitration cases preserving bargaining unit work from management takeover" (Kohl, 1982, 67–68).

The International Association of Machinists cites a case in which management attempted to disguise the introduction of new product lines as "development work" and assigned it outside the bargaining unit to a special area in or near the plant. When a complex precision component was brought into the main plant, it was given to a group of low-skilled contract workers and a group of "lab-technicians" holding two-year higher-education certificates—all outside the bargaining unit. After a long struggle, the union succeeded in drawing the jobs into the unit. In another case, when the company introduced an automatic press, management sought to keep union workers off part of the job, on the grounds that its control system was operated by a tape program. The machinists were "ordered not to do any programming but to work cooperatively with non-union programmers" (Nulty, 1982, 124–29).

As clerical procedures have been computerized on the railroads, management has tried to transfer the new jobs out of the bargaining unit. A dispute over just such an issue triggered an eighty-two-day strike at a major railroad in 1978 (see Chamot and Baggett, 1979, 33).

Concern about the fragmentation of bargaining units has led many unions to insist upon agreement clauses limiting the contracting out of work. For example, the Steelworkers agreement with United States Steel provides that before the company can "contract out a significant item of work," it will notify the union. If the parties cannot adjust any dispute arising out of the proposed shift, it will go to the grievance procedure and, if necessary, to final and binding arbitration (U.S. Steel–United Steelworkers, 1980, 14–15). Contracting-out clauses such as these were not directed only at technological change problems, but as such change has intensified, these clauses have been invoked more and more to deal with these problems.

Older Wage and Classification Practices

Some companies have taken the opportunity to extract special concessions as a price for investing in and installing new technology in older, potentially obsolete sites. At Lynn, Massachusetts, in return for building a new, highly automated "factory of the future," General Electric won from the union a separate contract to cover this special installation. Included among the terms were the right of the company (on four weeks' notice) to install twelve-hour work shifts (with possible consequent losses of Saturday and Sunday premiums for workers as well as loss of pay for jury duty or bereavement); the installation of measured day work; and the substitution of only three job-wage classifications (with greater flexibility of deployment of the work force by the company) instead of several dozen previously provided under arrangements between GE and the International Union of Electrical, Electronics, Technical, Salaried and Machine Workers (IUE). With the new work hours, employees would alternate between one week of four days and a second week of three days. Union officials claim that this could result in a 20 percent increase in annual earnings, presumably from a combination of more weekly hours and overtime premiums. In negotiations with the union, the company clearly warned that it would build the proposed new factory in an entirely different location if the concessions were not forthcoming (conversation with IUE staff official; also see IUE Local 201, 1984, 1985).[4]

The three classifications at the new Lynn plant will be repair control, automated factory mechanic, and stager. The switch away from the conventional skilled classifications—such as electrician, plumber, and so on—is especially significant and is part of a trend in many new computerized and automated factory operations. Many of the service and maintenance tasks in these new factories seem to require more general skills—combinations of some parts of the older, more highly skilled crafts. These new, general types of technical skill also tend to be more company- and plant-specific than was the case with the older crafts (Office of Technology Assessment, 1984, 110–112).[5] If such changes become widely extended in U.S. industry, they could pose severe adjustment problems for some of the traditional craft unions.

Another trend appearing in high technology firms is a move away from incentive and piece rate systems. As a representative of the IBEW remarked in conversation, this change has been occurring in some plants of Western Electric recently, since the company has been under pressure to preserve and advance product quality in the production of some delicate electronic goods.

The still-developing relationship between the United Automobile Workers and the New United Motor Manufacturing Company plant in Fremont, California, illustrates the impact of new production and organizing methods on traditional wage classifications. The company, a wholly owned joint venture of General Motors and Toyota, encountered trouble at its outset when Toyota (which holds the bulk of the direct managerial responsibility) balked at hiring from the laid-off worker list of the plant, which had originally been part of GM operations. Toyota also gave indications of its desire to operate without unions.

Under a compromise arrangement, the management, on reopening, accepted the UAW under a new, separate (from the GM master) collective agreement. Workers were rehired from the old layoff list, but some flexibility of choice was given to management, and old seniority rights were not recognized. (It does appear that the new management was interested in flexibility of deployment, and threats of operating nonunion were a tactical move to gain this prime objective. It is hardly plausible that GM could, even in partnership, operate a reopened plant on a nonunion basis, given the strength of the UAW at its other operations.)

The new agreement contains only four job classifications—three skilled and one production classification. Clearly, the company has a kind of flexibility of deployment of workers that exceeds the flexibility under traditional UAW–major auto company agreements, which contain many separate semi-skilled classifications. According to one report, production workers "are doing work performed by engineers and other salaried personnel at traditional car plants" as well as routine maintenance. UAW local officials at Fremont are clearly uneasy with all these changes and argue that "members want to be paid for these [new] tasks" (Bureau of National Affairs, 1985b, C3).[6] If these new methods of production, forms of work organization, and wage classifications offer advantages to management, it can be assumed that General Motors, Ford, and Chrysler will press the UAW to incorporate many of the same practices. Some experimentation with a few similar devices has already been attempted in a handful of older auto plants.

How sweeping the influence of rapid technological change can be on work and wage structures is also revealed in the experience of the CWA with AT&T in recent years. Union officials found that the "diverse and often complex . . . high technology workplace" of the telephone industry was resulting in the creation of numerous wage inequities, the existence of many obsolete job classifications, shifting responsibilities and skill requirements, the elimination of thousands of operators' jobs, and great undervaluation of numerous clerical jobs (Straw and Foged, 1982).

The union took a series of steps in its subsequent negotiations with AT&T to adjust some of these inequities. One major new union approach marked a departure from its traditional "apprehension and skepticism if not open condemnation" of job evaluation (Straw and Foged, 1982, 26). Cooperating with the company, the union appointed a joint Occupational Job Evaluation Committee, which is helping to implement a new jointly agreed upon system of job evaluation. The new job agreement is also linked to joint efforts, such as training and retraining and income security programs. The union's "conversion" to job evaluation resembles a similar plunge taken by the United Steelworkers of America over forty years ago, when they undertook a similar venture with the major basic steel companies.

As technological change affects existing job classifications, unions may seek to protect the wages of affected employees. "Red-circling," or the maintenance of the employees' older (if higher) wage, is often an objective of the union. An agreement in the newspaper industry states:

> [No] full time employee covered by this agreement . . . shall be dismissed or suffer any reduction in classification as a result of the introduction or use of automated processes, computer equipment or new mechanization. (Murphy, 1981, 16)

A June 1982 agreement between General Electric and the IUE includes a special provision aimed at employees whose jobs are "directly eliminated" by "a robot or an automated manufacturing machine." Provided that they have transfer rights to other jobs, they retain their old rates "for a period up to 26 weeks" (Office of Technology Assessment, 1983, 92). Of course, not all unions have been successful in negotiating these or similar provisions as yet, but they are indicative of a new trend.

Advance Notice of Technological Change

Clauses guaranteeing advance notice of significant technological change have always been a principal defense unions have sought in collective bargaining. Recent contract surveys have found that such clauses seem to be on the increase as the new technology bites more deeply into established industries.

Training and Education Programs in Collective Bargaining

Traditionally, craft unions, collaborating with employer groups, have assumed a major role in the training function for their occupations, especially in helping to administer apprenticeship programs. In most instances, industrial unions have traditionally left the training function to management (Barbash, 1963, 48).

Recently, the heavy impact of technological change has altered the approaches of both groups.

The Graphic Arts International Union, a merger of previously independent craft unions, has established training centers around the country through which 3,000 employees pass "every semester." These centers use the latest technology in the printing industry—often obtained with the cooperation of the industry, which shares the interest in better trained workers. The GAIU feels that it now has "the tools to cope" with a "high technology industry" (Stagg, 1982, 163–64).

The greatest change in training and retraining practices has come on the industrial union side, as these unions seek to help displaced members or protect remaining members who need new skills to retain their often rapidly changing jobs. The CWA has negotiated contract provisions with AT&T that, in the face of an "environment of fast paced technological developments," provide not only "job specific training" programs, but also more generic training for employees' "personal career developments" and to fill "job vacancies as anticipated by the companies" (AT&T–CWA, 1983, 15–16).

Agreements involving the outlay of millions of dollars have been negotiated in the automobile industry by the UAW to assist workers laid off for economic reasons and as the result of technological displacement. At the Ford Motor Company, the 1982 collective agreement provides an Employee Development and Training Program, administered through a separately established and jointly governed National Development and Training Center.[7] The program includes personal counseling and development assistance, such as basic literacy and communication skills and computer training and targeted training (often by specially hired outside specialists or schools) for a particular location or segment of the work force. The new center is also working out a wide-ranging college and university options program, which "will provide tailored college-level business and technical education programs to meet" the career needs of employees. Under this program, by agreement with many colleges around Ford plants in a number of states, training can be provided in the plant "and credit obtained for certain work experience." Employees can receive tuition and fee credit (in advance) up to $1,500 per year. Through 1984, some 6,200 employees had taken advantage of such grant opportunities.

For displaced employees, similar counseling and training programs are also operating, as well as professional placement assistance. In two plant shutdowns, 70 percent of the employees were helped to find new jobs within a year (for the most part, at comparable wages), according to union sources. A large number of these employees were retrained, often in specially tailored programs contracted out to professional trainers.

Where new technology hits hard, a common concern of employees is their ability to advance with it. The assurance, through a collective agreement, of adequate training opportunities for such employees can be a major appeal of unionism in the high tech age. One can look to a growing demand for union participation in such training programs.[8]

Automation in the Office and Health Issues

Unions with a large white-collar constituency have taken the lead in spotlighting the health problems posed by new electronic devices (see, for example, AFSCME, 1983). They have pointed, for instance, to the greater stress experienced by secretaries working with video display terminals. In addition, the 9 to 5 section of the Service Employees Union has argued that new electronic equipment is being used to subdivide some office work to the point that the office resembles a piecework factory. The result is a deskilling process and the creation of boring, unsatisfactory jobs, with greater stress and mental fatigue. Radiation exposure levels from VDTs are also a major concern (Bureau of National Affairs, 1984c, 18–19).

Responding to the problems of VDT users, the American Newspaper Guild (ANG) has formulated a mandatory bargaining program providing for the transfer of pregnant employees from hazardous working conditions, protection against low-frequency radiation by the use of better shields and other devices, periodic eye examinations for workers using VDTs, payment of the cost of glasses or contact lenses needed for working with VDTs, hourly breaks, and no use of VDTs during the final hour of a shift (to allow employees' eyes to adjust to normal light). Several of these provisions or variants of them have been incorporated in a recent agreement between the ANG and the magazine publisher Time, Inc. (Bureau of National Affairs, 1984c, 18–19).

The Service Employees Union recently negotiated a breakthrough agreement with one office of the Equitable Life Assurance Society in Syracuse, New York. It provides employees "extra break time . . . as well as the right to request transfers to non-VDT work if they are pregnant, or to other terminals if they believe their own to be unsafe." Safety equipment, including glare-reduction devices and detachable keyboards, are also required (Bureau of National Affairs, 1984c).

In addition to health problems, there is the potential for invasion of a worker's privacy by companies using remote control computer monitoring to check workers' output and perhaps setting new piece rates or production standards based on such monitoring. Two CWA officials report that the union has, by agreements, limited telephone companies' right to "monitor and control workers from centralized locations" or to otherwise use electronic means to rob workers "of autonomy" and "dignity" (Straw and Foged, 1983, 166). In a vein similar to the CWA agreements, the IUE has warned its local unions to take steps to "prohibit the surveillance of individual workers by television cameras" as well as by computer-based monitoring devices. It adds that "cameras connected to robots to help perform their work must be limited to that function."[9]

As yet, these types of agreement clauses and actions are the exception rather than the rule in U.S. bargaining. In Western Europe, several unions have been able to negotiate such safeguards for whole industries in special new-technology national agreements with employers' federations.[10] The lack of a predominant white-collar union in the private sector and the more fragmented structure of collective bargaining make progress in this area harder to accomplish in the United States. The labor–management system today is different from what it was in the past, when powerful auto and steel unions established bargaining patterns that could set the standard for much of the U.S. labor market.

Job and Income Security against Technological Displacement

For decades, unions have striven to protect their members from the adverse effects of economic and/or technological change. In many cases, of course, it is impossible to separate the economic-market causes of displacement from the technological causes, and protection clauses in collective agreements often do not distinguish between them. The use of attrition formulas, early retirement, transfer rights, and income guarantees—including severance pay and supplementary unemployment benefits—has become common in collective bargaining, and many unions have placed even greater emphasis upon them.

The typographical union at the *New York Times* negotiated sweeping protections in 1974. The union conceded to the newspaper the right to have reporters prepare "a news article on a computer terminal keyboard that could be used to operate an automatic typesetting machine." This change eliminated craft worker typesetting, but the newspaper "guaranteed lifetime jobs to the printing craft workers." Some workers chose early retirement, some accepted retraining for other jobs on the paper, and others accepted large lump-sum severance benefits payments (U.S. Bureau of Labor Statistics, 1982, 13).

During the course of the 1982 negotiations with General Motors and Ford, the UAW, in return for economic concessions on wages and holidays, obtained from the companies some new guarantees against plant shutdowns and some contract language that limited out-sourcing of automobile parts and assemblies. The same agreements also set up experimental programs at a handful of plants that provided new employment guarantees designed to furnish "life-time job security" for "80 percent of the workforce" at each of these facilities (Bureau of National Affairs, 1982).[11]

In 1984, at General Motors and Ford, the parties again bargained over the problems of displacement arising from technological change or out-sourcing. In regard to the latter, the companies have pressed to shift more production to facilities overseas, where labor costs are much lower. The new agreements are complex and can be summarized here only briefly.[12] A special job bank is created, which protects employees, at full wages and benefits, for a period of approximately six years from displacement due to four causes: technology, out-sourcing, negotiated productivity improvements, and job loss due to transfer or consolidation of work within the company (usually to a more efficient facility). Technology is "broadly defined. . . . as any change in product, method, process or the means of manufacturing at a location." Displacement due to new organization or new materials would be covered, and displacement due to changes in production would not be covered (Young, 1985, 456).

The jobs displaced as a result of these four factors are "deposited" in the job bank. Qualified individuals would receive their regular compensation for these jobs. If a more senior employee was later displaced for economic reasons, he could conceivably bump a worker out of the bank at his particular plant. On the other hand, if volume of production increases, laid-off workers might move into job bank slots. Job slots are eliminated from the bank by attrition.

Assignment in the bank means continued employment; it is not a mere unemployment or layoff income benefit. A worker in the bank can have a wide range of assignments, such as placement in a training program, or can replace other employees who go for training. At General Motors, a special account, not to exceed $1 billion, is set up to insure these benefits over the expected six-year life. At Ford, the account is not to exceed $280 million. One critic of the program comments: "In essence GM has paid a billion for an unlimited license to automate and out-source" (Shaiken, 1984).

A UAW view, expressed by its research director, holds that although the union has conceded to the two companies the right to make out-sourcing and new-technology decisions, the new "program does significantly increase the cost to the company of eliminating jobs as a result of such decisions." And "this should enhance significantly the priority of job security as a factor in auto company decision-making" (Friedman, 1985, 556).[13] It appears that the difficulty of adequately policing out-sourcing, as well as the companies' need to compete with Japanese and other companies, led the union to accept this compromise solution.

Many questions will arise in the application of this program, such as determining how many jobs will be directly eliminated by a particular technological change. These are to be solved by joint company–union committees at each plant and by joint area committees where necessary. A UAW official notes that reliance on such joint committees is "a continuation of an evolution" that goes back some years. They involve "a less legalistic approach to collective bargaining" and rely more on "good faith interpretation, rather than simply insisting upon any short term advantage" (Young, 1985, 457).

These 1984 auto agreements may have wide-ranging effects on other collective bargains that try to cope with technological change in the years ahead. Of course, the pace-setting character of auto (or steel) collective bargaining is more or less over; however, where unions are strong and industries are stagnating or declining, this approach to technological displacement may have appeal.[14]

The Future of Bargaining in High Tech

It seems fairly evident that no single union will dominate the labor bargaining side when and if high tech industry, even high tech electronics, is organized. A number of unions are already fairly well represented in the electrical machinery and electronics industries, including the IUE, UE, CWA, IBEW, UAW, and IAM. Pluralization among national unions has important ramifications for the type of collective bargaining that might develop when and if organization increases in the new electronics areas. Probably, the bargaining pattern established by the steel and auto unions in their respective industries, and the impact it has had on the labor market nationally, will not be repeated. The varied character of the electronics industry and its firms, as well as the multiplicity of unions, will encourage more localized forms of bargaining.

Useful as it has been to pick out highlights of particular new collective bargaining developments that reflect the impact of technological change, these parts may not sufficiently reveal the influence of the whole. This is particularly true when technological factors are linked to other developments, such as competition from foreign and nonunion sources as well as change in patterns of consumption.

The Status and Prospects of Union Organization in High Tech

As the AFL-CIO moves to recover some of its lost ground, high tech is just one among a number of difficult industrial sectors. Admittedly, high tech's lead position in innovation and the wide publicity it receives may make it strategically more important for organized labor's image than some other sectors. It is extremely difficult to make estimates of the extent of existing union organization in high tech industry. We do know that in certain broad industry categories— such as electrical machinery manufacturing, which includes the electronics industry—the approximate unionization rate of 21 percent of all wage and salary employees in the industry is far behind the unionization rates in well-organized sectors such as autos (59 percent) or basic metal production, including steel (48 percent); the chemical industry has a unionization rate of about 18 percent.[15]

Another way to look at the high tech unionization issue is via occupational groups in the population. In 1984, although one-third of all semiskilled manual operatives and about 30 percent of skilled craft workers were union members, only 16.1 percent of managerial and professional workers were unionized. For professionals, the unionization rate was 23 percent, whereas for technical workers, the rate was 12 percent. About half of the organized professionals, however, were teachers (the bulk of whom were in the independent National Education Association, with the AFL-CIO's American Federation of Teachers second). The number of teachers in labor organizations, according to a 1980 survey, was 1,688,000, or almost 62 percent of all teachers. Union engineers numbered 108,000, a little over 10 percent of all those employed. Unionized engineering and science technicians numbered 109,000, or 22.7 percent of all such employees. Professionals and technicians were fairly well organized in transportation, public utilities, and government (including teaching), but only about 11.3 percent were organized in manufacturing. Indeed, in the entire private sector, the number of organized professional and technical employees was around 10 percent in 1980 (Adams, 1985; U.S. Bureau of Labor Statistics, 1981, 8–11, 26–29).[16]

Unionization of Engineers

Professionals and technicians are projected to show the greatest proportional employment increases in the decade or so ahead. Within this group, professional engineers and engineering technicians, key occupations in electronics and other high tech industries, are expected to undergo especially great employment expansion (Silvestri, Lukasiewicz, and Einstein, 1983). This poses a special problem for union organizing efforts, since until now there has been only limited success in unionizing engineers.

The issue of unionizing large numbers of professional engineers and scientists has been before the AFL-CIO on and off for several decades. Periodically, new approaches have been made to these groups, which are urged to recognize that their status is like that of other employees and that they therefore need unionization. Recent surveys of engineers in the Massachusetts high tech area, however, indicate relatively low levels of interest in unions and a generally conservative bent among the engineers (see, for example, Chamot, 1978).

Thousands of engineers are organized in independent associations, especially in the aerospace industry, TVA, and a few other sectors. From time to time, several of the independently unionized associations have flirted with some AFL-CIO national unions. But either because association leaders could not carry their members or because the leaders were just using this flirtation as pressure on their employers, these moves have come to very little. A number of AFL-CIO unions have also organized some thousands of engineers scattered in a score of industries. A few unions, such as the IUE, have established special divisions to appeal to them.

Perhaps the independent engineering unions will align themselves into a tighter national association and get on with organizing engineers; but this seems doubtful, given their past experience and lack of necessary resources. In high tech areas such as Silicon Valley and Route 128 in Massachusetts, job hopping to take advantage of good markets makes the task of organizing engineers even more difficult.[17]

Obstacles and Responses to Unionization
in High Tech Industries

In electronics alone, high tech can mean anything from giants such as IBM or Western Electric (itself only a part of AT&T) to a small enterprise consisting of a handful of engineers and a few assistants. Moreover, within a clearly high tech firm such as Atari or Western Electric, the jobs to be done will vary from the most routine assembling or testing to the most advanced engineering research and design.

For purposes of illustration, however, let us restrict our discussion to a comparison of two clearly identifiable groups—the electronics and related high tech firms in such areas as Silicon Valley and along Route 128 in Massachusetts, and the giant national firms such as Western Electric, IBM, General Electric, and North American Rockwell. In both groups, one finds a two-tier work force. On one level are a substantial number of highly trained and highly paid engineers and technicians, often treated with kid gloves by management. Their wages and benefits usually put them near the top of the compensation ladder. At a lower level are a substantial number of low to moderately skilled manual workers (assemblers, testers, and so on) and a small cadre of skilled craftspeople. A smaller number of routine clerical employees rounds out the work force. Generally, the manual and clerical workers outnumber the professionals and technicians.

When we look at the wages and benefits of semi- and unskilled workers, we note a breakdown of the similarities between the two types of high tech companies. At Western Electric, for instance, the semiskilled operators and assemblers are paid anywhere from $8 to $11 per hour and have the usual array of fringe benefits enjoyed by strongly unionized workers, whereas in Silicon Valley, semi- and unskilled manual workers typically earn anywhere from a little over the minimum wage up to $5 or more per hour. Their fringe benefits are relatively meager.

The first obvious explanation for the greater advantages of the Western Electric employees is that they are relatively well unionized. But what of the nonunion, well-paid IBM manual workers? Perhaps the union substitution effect counts here (that is, IBM's willingness to maintain high wage and benefit standards to avert union organization), but this would appear to be only a partial explanation.

Firms such as Atari and most other high tech companies in Silicon Valley or Massachusetts appear to be caught up in a far more competitive atmosphere than Western Electric or IBM. The latter firms (along with a handful of other very large companies deeply into high tech production) are much more shielded economically. These giant companies are not subject to much competition, for a number of possible reasons. Superior product development or market factors may give them a semimonopoly position, or they may have a longer history and greater capitalization or inside tracks for important defense contracts; most probably, combinations of all these factors are important. In addition, product quality, especially in defense items, may play a larger role for the major companies; and they may pay a wage premium to ensure this quality. Part of the explanation for the greater foothold gained by unions in some of these larger companies is their ability to bargain and gain benefits in the more stable atmosphere that results from these factors.

There seems to be little or no unionization among the professional engineers in either group of companies. Western Electric engineers were fairly well unionized in an independent association several decades ago, but the usual problems of holding unionized engineers together, combined with effective company opposition, killed the union.

The poorly paid assembly and related semiskilled workers in many high tech companies would appear to be a likely target for unionization, especially in areas such as Silicon Valley or along Route 128 in Massachusetts. Relative to many other states, both states have good unionization levels and state labor laws that are better than most. But most of these firms are very mobile, in several senses of the word. They are often in and out of business rapidly. Product cycles within these firms move rapidly from development to production to obsolescence as a result of domestic or foreign competition, including new products that replace old ones at a phenomenal rate. Where firm life expectancy is short, any union investment in an organizing campaign becomes dubious.

Even when firms are larger and more durable, as in the Atari case, production operations at any particular location are often transitory. Most operations tend to be labor-intensive and involve relatively low fixed capital costs. James Parrott (1981) has provided a good description of the product cycle and labor force requirements in the electronics industries; the following three paragraphs are based on his work.

With low "fixed capital costs and high value to weight ratio of material inputs and product outputs," these firms "are heavily influenced by the availability of labor supplies in determining plant location." If the machinery is expensive (for example, "elaborate printed circuit board testing equipment"), it is usually "light in weight and relatively easy to move." The firms depend on two types of labor: one is the highly skilled professional force of engineers, scientists, and computer specialists; the other is the "relatively low-skilled workforce engaged in both assembling and testing operations in production and clerical function."

Once a product has been developed, the production requires less and less input from the professional workers. Indeed, once mass production is reached, location is no longer tied to electronics R&D centers "and can be shifted to take advantage of lower production labor costs." Such shifts are, of course, easier for larger firms with developed communications networks. Small firms have greater difficulty in moving.

These electronics firms, then, have tended to seek out areas such as Massachusetts or California's Silicon Valley, where there are pools of cheap labor for production and high tech professionals for product development. Competition in the electronics industry tends to be intense, but it appears that in the aggregate, the industry will have a healthy future, since it will be fed both by heavy defense orders and by the growing requirements of the civilian economy. (The aggregate, of course, includes high failure rates among many individual companies and among many divisions of even larger companies.)

In California, the industry draws its cheap labor from a pool of legal and illegal Latin and Asian immigrants as well as women entering the labor market and migrants from other states. The industry's postwar location in Massachusetts was influenced, of course, by the presence of MIT and Harvard; but the availability of a large, displaced, and relatively low-paid textile labor force was also important.

If anything, the casual way in which production labor is treated may be greater in Silicon Valley than along Route 128. In the latter, several longer-established companies pride themselves on stability of employment records.

The Atari Example

The problems encountered in the unionization attempt at Atari are a useful example of potential organizing difficulties.[18] Large numbers of manual employees seem to have been fairly ripe for unionization. Low wages, mediocre working conditions, and indifferent supervision were characteristic complaints of manual workers. Indeed, the unionization campaign had its origins in a strong, spontaneous approach to the union, the Brotherhood of Painters and Allied Trades, by a group of Atari employees who showed up at a nearby plastics plant where the union already had an organizing campaign under way.

Late in 1981, the targeted bargaining unit was composed of about 1,000 employees, concentrated largely in the Home Cassette Division (HCD) and the Coin Operated Games Division (COD); however, when the company announced its plans to hire an additional 1,200 employees in the original unit and several others, the campaign was expanded. Atari seems to have employed some of the usual anti-union tactics, such as unilaterally announcing a substantial wage increase, questioning employees suspected of being union activists, and so on. But the campaign, according to union estimates, continued to go forward.

The union was close to reaching the required number of signed cards for an NLRB petition when, in February 1983, the company exploded a bombshell by announcing that they would be moving the HCD and one other major operation that the union had also heavily targeted. The company added that it would be laying off 1,000 employees in the next few months. According to one newspaper report, Atari fired 600 production workers on four hours' notice. Amid some of the closings and reductions, the company also announced a new division to manufacture "smart telephones"; but no laid-off Atari employee would be eligible for hire there.

Heavy layoffs made the campaign at the Home Cassette Division increasingly difficult, and it was finally decided to petition the NLRB for an election at the COD in May 1983. Challenges over the cards and names of those actually employed in the proposed unit delayed the union, which filed a second petition in June 1983. Shortly after, a second wave of heavy layoffs occurred, fragmenting the proposed unit. When the union supported its petition in the unit with signed cards, the company challenged the appropriateness of the unit. After further delays, agreement was reached on unit determination, and the election was set for November 1983. The union was defeated by a vote of 143 to 29. From an original target of 1,000, later expanded to as many as 2,700, the unit was reduced to a couple of hundred people.

Much of the Atari experience is the typical routine misery of today's union campaigns against any company. What differentiates it is the enormous variation in unit size as the company moved significant operations elsewhere in the United States and overseas. This is likely to recur in some other campaigns, when and if they are made, to organize the manual workers in Silicon Valley or Massachusetts.

The High Tech Mystique

Aside from the great mobility of resources, there are several other distinctive characteristics of many high tech firms that make them more elusive in terms of unionization. There is often a kind of mystique about them that identifies them as the special leaders of modern, innovative industry. Much of this mystique has a "hype" quality, particularly as it relates to the manual employees in many of these firms, who are relatively poorly paid. How heavy this hype factor can be was illustrated when General Motors took over the Texas-based Electronics Data Systems Corporation (EDS). The white-collar force at GM was led to feel that the relatively tiny EDS was taking over GM, so great was the fanfare and promise ("Some GM People," 1984).[19]

The managerial styles of many high tech companies also make them harder for unions to penetrate. Many of these firms are still very new, and the sense of being a "start-up" often translates to organizational informality. Extravagant parties for company employees are still common in the industry. Profit-sharing plans are also common; and the sense of working in an industry where great wealth is just around the corner can be infectious among all employees— especially the professionals. For these employees particularly, many companies also allow considerable autonomy on the job. All of these practices tend to make the appeal of or need for unionism weaker.

James Fallows notes that among the Silicon Valley companies there is "an atmosphere different from the 'mature' manufacturing firms." The Valley firms are "more flexible" and "less concerned with the trappings of rank." Another writer reports that an "egalitarian and flexible structure" allows the electronics industry to "avoid the bureaucratic structure that is characteristic of most firms" (Howard, 1981, 21).

It is true that these more flexible and participative practices are often enjoyed only by the professional and technical employees, while production workers in Silicon Valley face low-wage, dead-end jobs, unskilled tedious work, and exposure to some of the most dangerous occupational health hazards in all of U.S. industry (Howard, 1981, 21). But it is the "high" side of this industry that obviously catches reporters' and other outsiders' eyes, adds to the mystique of the industry, and makes it harder to organize.

The special industry mystique is sometimes enhanced by the public relations and related programs of high tech companies. In Massachusetts, a number of firms in high tech research and manufacturing came together in 1977 and established the Massachusetts High Tech Council. The council has lobbied for property and auto excise tax reduction, a rental deduction for tenants, and other legislation—often against the positions of other business associations. It also campaigns for an increase in technical education in the state and strives to educate member companies in human resource management techniques and procedures (Massachusetts High Tech Council, 1984). On occasion, human resource management can be a code word for union avoidance or union minimization, though admittedly this need not be the case. The council itself is composed mostly of nonunion companies.

New Union Strategies

Most high tech firms are new, and they have entered the industrial scene at a time when being anti-union has become the norm for most U.S. management. It is not surprising that they generally accept that norm. The kind of activity and public relations practiced by many high tech company groups helps further insulate high tech firms from union efforts to organize them. The unions are shut out from the specially created high tech world.

To counter this strategy, a couple of unions in New England, the Communications Workers of America (AFL-CIO) and the United Electrical Workers (UE–Independent) have helped to support a High Tech Workers Network. It has issued newsletters and has monitored developments in wages and working conditions in high tech companies. Cautiously, the group has disavowed being an organizing committee and speaks of winning improvements in how workers are treated on the job, whether or not enough support exists for collective bargaining through a union (Early and Wilson, 1983). In its own newsletters, however, the network describes the organizing efforts under way at several high tech companies. The network's newsletter devotes considerable space to probing the effects of the dangerous chemicals to which high tech workers are being exposed and, along with other unions in New England, has cooperated with environmental and safety groups critical of high tech companies.[20]

A group of AFL-CIO unions has formed a loose coalition to try to organize Massachusetts high tech companies. They have talked about dividing targets and carrying out joint education programs, and they, too, have cooperated with environmental and safety groups in lobbying for occupational health legislation. In 1983, they helped bring about passage of the state's "right to know" law about the dangers of chemical and related risks at work ("Right to Know," 1983). The high tech electronics work environment presents serious health dangers for production workers, since workers are exposed to carcinogenic substances such as solvents, acids, and fiberglass materials in the fabrication and testing of silicon chips (Pacific Studies Center, 1977, 32).

Union organization of high tech companies remains uncertain. It cannot be separated from the more general organizing problems that confront the AFL-CIO and its affiliated unions. Much of the AFL-CIO's new emphasis and proposed tactics for organizing have as much application to high tech (taking Route 128 or Silicon Valley as examples) as to other sectors. Given the often small or modest size of these companies, their large numbers, and their shuttling in and out of business, the AFL-CIO calls for greater emphasis on choosing appropriate targets and on closer cooperation in establishing a mechanism for resolving organizing disputes among affiliated unions in order to prevent wasteful organizing competition (AFL-CIO, 1985, 28, 31). There is always the danger, of course, that such forms of cooperation may reserve parts of the organizing jurisdiction to unions that have only "paper claims" to some organizing areas and lack either the will or resources to undertake such organization.

The large number of unions that have already carved out small bits of the high tech electronics area will probably make future organizing in this sector more difficult. The snowball effect that helped sweep thousands of steel, auto, rubber, and other workers into those industrial unions, once the unions had gained a head of steam, is not likely to be repeated. The multiplicity of unions in the field will hamper the development of a similar effect among high tech workers.

The high tech sector's spirit of buoyancy and optimism, its tendency toward mobility, its frequent use of participative management and profit sharing, and the union obstacles that I have outlined may indeed make it somewhat more difficult to organize than some other sectors. When and if more stability comes to the industry, however, the poorly paid high tech production workers make it an inviting union target.

Conclusions

The AFL-CIO and many of its affiliated unions take a somewhat pessimistic view of the new "technological revolution." They fear large-scale displacement of the labor force. But even stronger is the fear that this new technology may lead to a labor force in which there will be a relatively modest number of well-paid professionals and technicians on top, with a large number of low-paid, routinized jobs at the bottom. The concentration of AFL-CIO members in the "older" manufacturing industries—such as steel, autos, and rubber—where employment has been shrinking, will further reduce jobs and union membership.

Most economists are arguing that the current changes in new technology are not unlike past changes. Moreover, as the rate of labor force increase declines, by the late 1980s, some economists expect that improved capital/labor ratios in the service industries will provide a foundation for rising productivity, jobs, and wages in that sector.

There are some signs of change in the AFL-CIO positions as the labor federation adjusts to the new technology and to a different labor force. These changes include greater acceptance of quality of work life programs, other forms of worker participation and cooperation with management, and a less adversarial approach to bargaining. It is not always easy for unions to shift to less adversarial forms in the face of the continuing refusal of most managements to accept union recognition or existence. Many employers have sought to take advantage of changes in technology by shifting new work outside existing bargaining units.

In the past, craft unions seemed to be more affected by technological change; but in the recent technological revolution, industrial unions have been highly affected, and many of them are displaying a kind of "scarcity consciousness" that scholars had earlier identified primarily with craft unionism. Moreover, the current technology is radically altering many forms of white-collar work, evoking new responses by white-collar unions.

Many existing wage and job classifications are undergoing extensive change. Companies are seeking greater flexibility of deployment of the work force to take advantage of new technology and production organization; and a number of unions have yielded to these requests. To protect jobs, unions make use of older and tried devices such as demanding limits on contracting out, "red-circling" the wages of affected employees, requiring advance notices of technological change, and demanding transfer rights for displaced employees.

New technology often creates a need for training and retraining the work force, and this has led some industrial unions to take a major role in new joint educational programs with management. Craft unions, too, are expanding their role in worker training programs. More unions are likely to include jointly administered training programs in their future bargaining demands.

Unions have taken up the cause of gaining protection from some of the health and safety hazards resulting from new office technology such as VDTs. Unions have also been concerned about threats to workers' privacy that are posed by new automatic monitoring devices.

New income and job security programs, such as the job banks at GM and Ford and the employment-income security agreement for typographers at the *New York Times*, have been negotiated by unions to cope with new technological displacement.

There should be no exaggeration of the numbers of employees in the high tech sectors, especially as this relates to the general organizing tasks confronting the AFL-CIO. No single union will emerge as the dominant labor bargaining force when and if organization comes to high tech, even in the electronics field, since many unions lay some claim to the jurisdiction. Consequently, even if high tech areas become well organized, bargaining patterns in those industries are not likely to have the same general labor market impact as was the case after the auto and steel industries were first organized.

Despite the organization of some employees in a few major high tech companies such as Western Electric, the present general state of organization among high tech industries and occupations is meager, and attempts to increase it face considerable obstacles. Organizing of engineers, a key factor in many high tech firms, has been a problem for the AFL-CIO and some of its affiliates for many decades. The exploration of several possible alternative paths to organizing engineers suggests no easy breakthroughs. This is particularly likely to be the case in such high tech areas as Silicon Valley and Route 128 in Massachusetts, where the strong market for engineers makes job hopping very common and unionization difficult.

The labor force in many high tech companies, especially in electronics, is sharply divided between a large corps of skilled, well-paid professional and technical workers and semiskilled or unskilled, poorly paid manual and clerical workers (although some large companies, such as Western Electric and IBM, do pay their production employees well). The production employees in most of these firms, often with low wages and benefits and mediocre working conditions, might be a good union organizing target. The great mobility of firms, though often with high failure rates, continues to make organization difficult and union investment in such efforts dubious.

The great "hype" and public relations about the high tech industry helps insulate many firms from unionization. The management style in these companies—with emphasis on employee participation, profit sharing, company dinners, and other perquisites—contributes to the difficulty of organizing these firms. Many of these firms came into being during the past decade, when deep resistance to unionism had already become common among much U.S. management.

Unions have made some special efforts to counter adverse forces with joint organizing approaches and cooperation with environmental and safety groups that call attention to some of the serious health hazards confronting many production workers engaged in silicon chip fabrication.

The large number of competing unions that have already staked out claims to organizing in the high tech electronics industry will probably make future organizing difficult. There is not likely to be the snowball effect that occurred in the auto, steel, or rubber industries when the industrial unions in those sectors developed. However, when and if greater stability comes to the industry, poorly paid production workers could be a likely target for unions.

Notes

1. See, for example, the views of Nathan Rosenberg and Lawrence Klein in U.S. Congress (1983).

2. Nobel Prize winner Wassily Leontief is among the relatively few economists who believe the present wave is qualitatively different from the past and that it could lead to massive unemployment. He argues that millions of Americans will have to depend upon vast new income transfers for a decent existence by the year 2000. The traditional market mechanism can no longer be expected to bring "a reasonable satisfactory distribution of income," as "the new technology diminishes the role of human labor in production" (see Leontief and Duchin, 1984, 8; Leontief, 1983, 404).

3. See, for example, the articles by William Winpisinger, Tom Brooks, and William Gomberg on quality of work life issues in the AFL-CIO's monthly, *The American Federationist* (1972, 1973).

4. As a price for reaching agreement with the UAW to build the new Saturn model in the United States, General Motors won union acceptance that the plant would be covered under separate contract negotiations, which involves departures from past arrangements between the company and the union, including fewer job classifications and greater managerial flexibility in deployment of the work force.

5. In the GE-Lynn case, the repair control classification calls for, as minimum qualification, an "electrician's license and successful completion of a formalized company training program." For automated factory mechanic, minimum qualifications are "machinist's background" and experience in one stipulated company classification (IUE Local 201, 1985).

6. I realize that many of these changes are not strictly technological, but they do reflect new (for U.S. auto companies) production and work organization.

7. This and the following discussions of Ford–UAW is from literature on the new joint program and from an interview with UAW representatives on the joint governing

body. Important new funding for the Ford and GM training programs was added in the 1984 agreements.

8. A biennial survey of approximately 100 major collective bargaining agreements, completed by the AFL-CIO's Industrial Union Department in November 1984, indicates a striking increase (over its January 1982 survey) in the number of agreements that provide training rights for new jobs to displaced workers.

9. Resolution 61, on "robotization," passed at the IUE 1980 convention (mimeographed). A similar resolution was passed by the 1982 convention.

10. See, for example, a summary of the March 1982 agreement signed between the Norwegian central union federation, LO, and the country's employers' federation ("Norway," 1982). Similar safeguards on VDTs are incorporated in the agreements between the German Metal Workers Federation (IG Metall) and employer groups in German metal manufacture ("West Germany," 1981).

11. In some of the proposed locations, workers rejected these "experiments" because they excluded 20 percent of the work force, or because they could lead to transfers to Ford or GM plants that were much farther away.

12. See "UAW-GM Report" (Bureau of National Affairs, 1984b) for an official, detailed summary. (For treatment of displaced workers, the agreement with Ford is similar). Useful summaries and explanations of the 1984 contract negotiation results can also be found in Young (1985) and Friedman (1985).

13. I can recall similar statements about making job security a higher priority for auto companies by imposing income charges on them, made by a UAW research director when supplementary unemployment benefits (SUB) were first negotiated in the auto industry three decades ago. Clearly, despite some of the problems SUB funds have encountered, they have provided considerable income security for tens of thousands of workers.

14. These auto agreements bear some resemblance to the special automation funds established in 1959 to eliminate restrictive work practices that hindered productivity in the West Coast longshore industry (see Kennedy, 1962, 80–82). However, the great differences between the industries, as well as many of the new provisions in the 1984 UAW agreements, limit such comparisons.

15. Estimates are from unpublished surveys of the U.S. Bureau of Labor Statistics.

16. Labor organizations comprise a broader category than unions and include the NEA and some other professional associations. I have tended to use the terms interchangeably for the limited survey here. The U.S. Bureau of Labor Statistics (1981) study also distinguishes, in some of its tables, between employees actually in labor organizations and those represented by labor organizations, the latter figure generally being higher. I have avoided this distinction and have tried to adjust figures to a uniform, unionized basis, rounding figures at times. (The figures for members in the Bureau of Labor Statistics Bulletin 2105 are about 10 percent below those represented.)

17. According to one source, among "technologists" in Silicon Valley, "job hopping has been so common that the industry's annual turnover rate is 23 percent" ("The Computer Era," 1980).

18. This and the following few paragraphs on Atari are based on conversations with various Painter's Union officials who were directly involved in the unionization campaign, and on reports prepared by union staff involved in the experience. Newspaper clippings from several sources were also helpful.

19. EDS took over GM's data processing and 10,000 of GM's white-collar jobs.

20. See, for example, descriptions of such efforts at Digital Equipment Corporation and criticism of union avoidance efforts by management at Orion Research (Cambridge) and Honeywell Electro Optics (Lexington) (see *High Tech Workers Monitor*, 1983).

10
Human Resource Management and Business Life Cycles: Some Preliminary Propositions

Thomas A. Kochan
John B. Chalykoff

Interest in human resource management has blossomed in recent years at the same time U.S. high technology industries have enjoyed rapid growth. Because many of the most successful and visible high technology firms have earned reputations as innovators in human resource policy and practice, it is understandable why some might confuse the *association* of high technology and human resource innovation with a cause-and-effect relationship. The purpose of this chapter is to suggest that such a view is superficial because it fails to recognize the basic forces that stimulate and maintain innovative human resource policies and practices. To appreciate the underlying forces, we need to develop a more systematic theory of human resource policy. With the help of a stronger theoretical foundation, more grounded predictions can be made about human resource management both in the high technology sector and in firms operating in other parts of the economy.

In no field of the social and behavioral sciences other than human resource management has organizational practice, management consulting, prescriptive academic and textbook writing, or teaching so far outstripped theory and hard empirical evidence. The lack of a sound theoretical foundation in the face of a growing demand for information by practitioners has led to the proliferation of popular literature that focuses on process rather than content issues. For example, managers are constantly bombarded with the obvious and difficult-to-refute advice that effective human resource management policies require human resource planning, which, in turn, requires effective integration with an organization's strategic planning processes. If this is true, then we have both theoretical and empirical grounds to predict that high technology firms have very ineffective human resource policies. This is because strategic planning is most difficult to do and least likely to actually determine policies in environments or organizations that are changing rapidly (Mintzberg, 1985). A recent study of human resource planning in a sample of high technology firms located in the Boston area concluded, for example, that products, markets, and organizational structures were changing so fast that few of the firms studied had human resource planning systems that came close to fitting the prescriptive models found in the literature (Wilson, 1982). Of course the author of that study has a prescription: These firms need to remedy this deficiency by developing more systematic planning processes linked to their business strategies. The point of this example is not to suggest that the advice about planning is wrong but to demonstrate the need for a deeper conceptual base to explain and evaluate the practices that are unfolding before jumping to prescriptive statement.

Tasks of a Theory of Human Resource Management

What, then, should go into a theory of modern human resource management systems? Although a comprehensive theory and research program will require addressing the variety of questions concerning the antecedents and consequences of human resource management policies (Dyer, 1984), we want to focus here on a narrower set of questions—namely, what accounts for the *development* and *maintenance* of human resource policy innovations that add up to what might be labeled a new human resource management *system*. We see this as an initial step toward development of a more analytical and testable theory of the substantive content or features of this system, their determinants and dynamics, and ultimately their effects on the goals and interests of the parties to employment relationships. First, then, we need to outline the substantive dimensions of such a comprehensive new system.

Table 10–1 presents several broad dimensions of what our larger research group (see Kochan, McKersie, and Katz, 1985) has identified as a new set of comprehensive human resource policies. The list is not meant to be exhaustive. Rather, it seeks to focus attention on activities at three levels within the firm: (1) the top level of organizational decision making, where the basic beliefs and value premises are formed that guide organizational human resource policies and the business strategies of the firm; (2) the middle or functional tier, where human resource staff professionals, line managers, and, where present, trade union leaders traditionally have interacted to establish and administer specific personnel or industrial relations policies; and (3) the work place level, where individuals and work groups interact with supervisors, union representatives, and each other in implementing these policies and experiencing their effects.

This three-tiered framework is drawn from our group's larger effort to reconceptualize the nature of industrial relations systems at the level of the firm

Table 10–1
Comparisons of New and Traditional Human Resource Management Policies

Firm Level	New System	Traditional System
Strategic	Active influence of top executives in human resource management policy development	Human resource policy recommendations developed by staff and approved by top executives
	Human resource management staff participates in key business decisions	Personnel or IR policies managed in a defensive or reactive fashion
	Human resource management planning process linked to strategic plans	
Functional	Contingent compensation practices	Wage policy driven by external standards
	Employment stabilization for core employees; flexible contracts for others	Layoffs used during short-term demand fluctuations
	Compensation and benefits set at or above labor market median	
	Promotion from within	Promotion and training policies a function of labor market conditions and contractual rules
	Extensive training and career development	
Work Place	Due process managed through multiple informal and nonbinding communications, appeals, and counseling processes	Due process through grievance procedure and binding arbitration
	Flexible work organization systems	Individual job assignments and rules
	Skill- or knowledge-based pay structure	
	High levels of individual and internal work group participation in task-related decisions	Little individual or group participation in task decisions

(Kochan, McKersie, and Katz, 1985). In that work, we add to the framework parallel levels of policies and practices for labor unions and the government, since these are two additional actors affecting human resource management or industrial relations activities, either through direct interactions with the firm or by the threat of potential interactions. Thus, any theory of human resource management should be integrated into a larger model of the industrial relations context or system. This serves as the first central premise of the model developed here.

A second premise of this approach is that there are systematic patterns or interrelationships among the policies and practices that cut across the three levels of human resource management activity. It is this pattern that can be said to constitute a human resource management system. More specifically, these interrelationships define the coherency, or lack thereof, of the overall human resource management system. Coherent human resource management systems have human resource policies that are consistent across all three levels of human resource management activities.

The dimensions of this new human resource management model outlined in table 10–1 can be summarized briefly. At the highest tier of activity, the policy is generally supported by a strong commitment from the chief executive of the organization to values that are conducive to giving human resource management concerns a high priority in organizational decision making. In much of the popular literature (Peters and Waterman, 1982), the chief executive officer's role has been overly idealized and sanctified and has been taken as the dominant or sole driving force in the development of new policies of managing people. Less attention is given to the deeper question of why the role of top management values and beliefs has surfaced as such a visible and seemingly important causal force at this particular moment in our historical experience. Out of this current "cult of the executive" has arisen a large literature on organizational cultures and their effects on human resource management policies. It is believed that a strong culture—initiated and nurtured by the beliefs, values, and actions of the chief executive—can filter down throughout the organization and can be sustained over time in ways that condition the alternatives considered acceptable or unacceptable in making decisions.

Although we believe that the organizational culture literature grossly overstates the centrality of the chief executive, it is also clear from a good deal of literature on organizational change that the values and the commitment of top decision makers in an organization are critical to the acceptance and nurturing of a comprehensive new policy. Thus, it is not surprising that we see the need for chief executives with values attuned to or committed to a strong role for human resource policy as a key dimension of these policies at the strategic level of decision making within the firm.

A second ingredient at the top level of policymaking is the existence of a human resource staff that has an influential role in executive decision making. In recent years there have been growing signs that human resource executives have increased their influence within organizations and that they have been moving into higher levels of executive decision making. A closely related third component at the strategic level is a feature criticized at the outset of this chapter—namely, the existence of a human resource planning process that does attempt to link the organization's human resource policies to the business strategies of the firm. In some cases, the firm may even adapt its business strategies at the margin to fit its human resource policies.

At the middle tier of the framework, firms following the most innovative human resource policies tend to have compensation and fringe benefit levels that are at or above the median in the labor markets in which they compete. In addition, compensation policies tend to be more contingent in design. That is, salary increases are more closely tied to the economic performance of the firm, and a higher percentage of total employee income may be derived from profit sharing, bonuses, or merit adjustments and differentials.

Staffing policies of firms following the new human resource management model reflect a commitment to employment stability for a core set of workers, supported by sets of less secure part-time or temporary employees, subcontractors, or other buffering agents able to absorb fluctuations in product demand. Efforts to avoid layoffs for core workers in response to short-term fluctuations in product demand also lead to reliance on work sharing, wage and salary freezes during cyclical downturns, and well-developed transfer and retraining systems for redeploying labor during periods of structural shifts in business strategies or product demand. Employment stabilization policies are further supported by a commitment to employee training, development, and career planning, along with a policy of emphasizing promotion from within rather than external recruitment.

At the work place, the policies that support and are most consistent with those described at the functional and strategic levels are those that maximize the dual objectives of flexibility in the utilization and deployment of workers and a high degree of commitment, loyalty, participation, and motivation on the part of individual employees and work groups (Walton, 1980). These dimensions of work organization and management of individuals are reinforced by communications programs and due process mechanisms for resolving conflicts involving individual rights, problems, and complaints. More specifically, our research on industrial relations has identified three interrelated dimensions generic to work place management of human resources (Katz, Kochan, and Weber, 1985; Kochan, McKersie, and Katz, 1985). The choices made concerning these dimensions largely define the substantive human resource policies at this level of the framework.

The first dimension has to do with the management of conflict and the provision of due process mechanisms at the work place. In the most innovative organizations, a variety of mechanisms—including grievance procedures, ombudsman services, meetings with managers to discuss problems, and other oral and written communications mechanisms—are used to achieve this purpose (Aram and Salipante, 1981; Balfour, 1984; Rowe and Baker, 1984). The second dimension involves the rules and principles for organizing work and designing jobs. The work organization system tends to emphasize relatively few job classifications, the use of work teams as opposed to individual work assignments, compensation systems that set pay on the basis of skill attained rather than the specific job performed on a given day, and flexible assignment of people to different tasks (Verma, 1985). The third dimension focuses on the extent to which the firm attempts to bring individuals into the decision-making process surrounding their job in an effort both to better motivate individuals and to improve productivity and product quality through more decentralized organizational decision making and communications. The most frequent terms for the processes used to carry out this objective are quality circles, quality of working life programs, employee involvement, or employee participation teams.

Although few organizations are likely to have all of these human resource management practices at these three levels, the policies listed in table 10–1 do summarize much of what both the popular and the more analytic literature suggest are characteristic of the most innovative human resource management organizations. Moreover, for our purposes, it is important to note that the policies found at the three levels fit together in a coherent fashion. That is, to develop and maintain over time the high level of trust and efficient and flexible work organization systems requires policies at the strategic level that give human resource management considerations a high priority when making business strategy decisions and adjustments. This, in turn, requires the support of top executive decision makers. Moreover, commitment to policies of employment stabilization requires flexibility in labor costs. Therefore, contingent forms of compensation fit better with an employment stabilization commitment than does a wage system that adjusts to fluctuations in product demand by adjusting the quantity rather than the price of labor. Organizations that attempt to implement a strong set of values of the chief executive, in turn, need to emphasize promotion from within so that employees can be socialized into accepting the values of the organization. Similarly, promotion from within provides additional reinforcement for a high level of employee commitment to the firm and an acceptance of profit sharing or other contingent forms of compensation.

This set of policies stands in contrast to the more traditional human resource management or industrial relations system that grew up and became institutionalized during an earlier era. This earlier system was based on the principles of scientific management; job control unionism, which emerged after the passage of the New Deal labor policies; and the job evaluation and other administrative principles that grew out of the personnel management profession of the 1930s and 1940s. At the strategic level, the traditional model had less integration between strategic business decision makers and the industrial relations or personnel staff. This was partly due to the need to respond to union pressures. It also reflected the lack of commitment of top executives to the role of unions in the firm and to the collective bargaining process. Instead, unions and collective bargaining principles were something that had to be tolerated because there was no viable strategy for union avoidance. Therefore, industrial relations staff were rewarded for stabilizing labor–management relations and for avoiding uncertainties that might inhibit the attainment of strategic business objectives. Industrial relations and personnel policies and the staff specialists responsible for them were therefore relegated to defensive or reactive roles, rather than active, planning, or policy-initiating roles.

At the functional level, wages were set in a more externally oriented fashion by emphasizing comparison of wages across competing firms in an industry and in response to changes in aggregate productivity and cost of living. Employment fluctuated more directly with fluctuations in product demand and business conditions. Layoffs were used as the chief mechanism for dealing with short-term needs to reduce employment costs, rather than work sharing or contingent or flexible compensation policies.

At the level of the work place, scientific management and job evaluation principles were incorporated into detailed collective bargaining contracts and personnel policy manuals for nonunion employees. Jobs were divided up along narrow lines, with specific wage rates, duties, and individual worker rights associated with each job. Over time, the number of job classifications multiplied and job descriptions narrowed. The major mechanism for assuring due process and conflict resolution under collective bargaining was the grievance procedure, with third-party arbitration. In nonunion settings, the personnel department was responsible for enforcing uniformity and equity in the day-to-day administration and management of these policies. Less emphasis was given to managing individuals or bringing individuals into decision making through direct employee participation.

It should be noted that the foregoing description of the traditional industrial relations system is also a very stylized ideal type. A great deal of diversity in personnel and industrial relations policies and practices existed across firms and industries operating under this system. Moreover, the traditional system was not limited to unionized establishments. Indeed, the literature on personnel management from the 1930s through the 1960s consisted largely of the analysis of techniques for implementing these same policies in nonunion work settings. During this period, however, new policies tended to first be adopted under collective bargaining and then to spill over to the nonunion sector—not only because nonunion firms wanted to remain unorganized but also because these practices generally fit well with the economic environment of the post-1930s period, the technology of mass production, and the prevailing theories of personnel management.

Determining Forces of Human Resource Management Innovation

A central conclusion of our larger research project on industrial relations is that over the past two decades, the new system outlined here has diffused and has become institutionalized in a large enough number of organizations so that it now serves as a significant competitive threat to firms operating under the traditional system. Because features of the new model have become accepted as prevailing practice or as state-of-the-art management policies, they are now being built into the design of most new organizations (Stinchcombe, 1965; Tolbert and Zucker, 1983). At the same time, however, we observe a growing pressure on some of the practices embodied in the new system. Thus, we need to understand both the factors that caused the new system to emerge where it did and those that will influence its future dynamics.

Several competing explanations for why new human resource management policies were adopted were implied in the foregoing discussion. In this section, we will make these explanations more explicit by offering specific propositions relating environmental changes, business strategies, management values, and organizational structures to different dimensions of the new system. Three broad premises guide the choice of variables and specific causal arguments presented in this section. First, no single causal force or variable dominates or provides a parsimonious explanation for all characteristics of the new system. Thus, we need to introduce the role of each of these variables as a causal force. Second, it is the *confluence* or *interaction* of these forces working in the same direction that accounts for the emergence of enough of the innovations across the three levels to produce a stable or enduring new human resource management system. Younger or newer high technology firms currently possess many of the necessary values, business strategies, and environmental conditions. This explains why high technology firms tend to exhibit many of the features of this new system.

A third important premise guiding development of the causal forces is that economic pressures to change or abandon some of these human resource management innovations increase as firms—or, more precisely, business units—move to advanced stages of their business or product life cycles or, for other reasons, face significant and sustained increases in price competition. As price competition increases, the confluence of forces giving rise to the emergence and stability of the new system breaks down, and firms experience internal debates among those who argue for maintaining the existing policies and those who propose changes that are believed to be necessary given the changed market circumstances. The outcome of these internal debates is indeterminate a priori and depends on the outcomes of internal organizational political processes (Mintzberg, 1985). What often results is a piecemeal adaptation (Katz and Sabel, 1985) which gives rise to internal contradictions or inconsistencies among policies at the different levels of the system (Kochan, McKersie, and Katz, 1985). Similar internal contradictions characterize the piecemeal efforts of many older firms operating in cost-competitive mature markets as they attempt to adopt only selected features of the new model. We will illustrate the types of internal contradictions that can develop by speculating on where the pressures on this new system are likely to be experienced as high technology firms move to advanced stages of their life cycles and as more mature firms attempt to adopt selected innovations in their search for new market niches to support growth strategies.

The key variables in this model and their interrelationships are diagrammed in figure 10–1. The model starts with environmental characteristics of (1) the labor market and the demographics of the work force, (2) government policies,

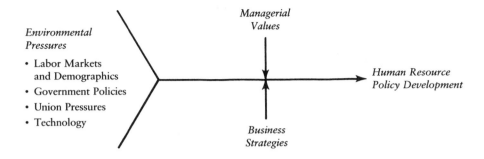

Figure 10–1. A Strategic Choice Model of Human Resource Policy Development

(3) the threat or presence of unions, and (4) the nature of technology. Business strategies and organizational values play mediating roles in determining how similar environmental pressures influence organizational policies in specific firms. These mediating variables introduce a range of discretion into firm policymaking and account for what we have labeled elsewhere a strategic choice model of industrial relations (Kochan and Cappelli, 1984; Kochan, McKersie, and Katz, 1985). Thus, the central arguments of this model are that it is the interaction of these environmental pressures and organizational choices that determines the content of human resource policies and that it is the degree of coherence or compatibility among policies found across the three levels of the system that contributes to its stability.

Environmental Forces

Labor Markets and Demographics.   Labor market forces and the demographic characteristics of workers provide a good starting point for the model because they exert the most exogeneous effects of all the variables to be introduced. Two propositions summarize the effects of these variables: first, the tighter the labor markets, the greater the human resource management innovation; second, the higher the proportion of the firm's labor force employed in high-skill, managerial, technical, and professional occupations, the greater the human resource management innovation.

Labor market pressures lie at the heart of both neoclassical economics and organizational efficiency explanations for the rise of internal labor markets and modern personnel practices (Hall, 1982; Jacoby, 1985; Baron, Dobbin, and Jennings, 1985). Although we agree with Jacoby and with Baron et al. that the efficiency view fails as a complete explanation of human resource innovation, tight labor markets clearly serve as an important driving force. Tight labor markets for critical technical skills are especially prevalent and pose major threats to the strategic business objectives of firms in the initial and early growth stages of business life cycles (Kochan and Barocci, 1985; Galbraith, 1985; Wils and Dyer, 1984). At this early stage, the firm (or business unit) often lacks the internal human resources and technical talent needed to transform into marketable products the ideas of the founder or the basic technological breakthrough that gave rise to the new business venture. It must recruit aggressively on the external labor market for technical talent in competition with other organizations that are working on similar new products or technologies. Thus, in the early stages of a business life cycle, recruitment of the best available talent is the critical functional human resource activity, since the ability of the business venture to exploit the growth opportunities available to it depends on its ability to get its new products to the market quickly (before its competitors). Human resource costs tend to be given lower priority. Furthermore, because time is critical, the firm prefers to recruit externally rather than to develop needed talent internally by investing in training or retraining of existing technical staff. This further drives up wage costs, since premiums have to be paid to lure scarce technical talent from other firms. Since turnover is costly to firms, extensive fringe benefit packages and, for high-level executives, incentives and bonuses are often used to deepen commitment to the firm.

These effects were observed in a case study we recently conducted of the selection system in a rapidly growing high technology firm engaged in defense contracting and computer software development (see Kochan and Barocci, 1985). The firm had several major government contracts in a newly emerging area of defense technology and needed to recruit a large number of electrical engineers and computer professionals with state-of-the-art training in order to meet the performance requirements of these contracts. Furthermore, the anticipated expansion of this area of the Department of Defense budget suggested that future contracts awaited those firms that could demonstrate their technical capabilities at this early stage of market development. Thus, recruitment of technical talent was the most critical human resource management activity in this firm at the time. The firm was prepared to devote whatever financial and staff resources were needed to get this job done well.

Although movements in the business cycle can produce tight labor markets at later stages of a business life cycle as well, this example illustrates how, at early stages, the competition for scarce talent (1) leads executives to give a high priority to attracting the talent needed to achieve business objectives and (2) creates incentives to shape a company culture that emphasizes the deep commitment of the firm to meeting employee needs, in the hope that top management concerns will be reciprocated with strong employee commitment and a low turnover rate. Employment stabilization policies may feature prominently in this environment, because the growth conditions needed to support such a policy are present and top executives want to demonstrate their concern for employee welfare in visible ways (Dyer, Foltman, and Milkovich, 1985). It is also clear in this environment why salary and fringe benefit costs can escalate rapidly for scarce talent and, in turn, can exert pressure to raise salaries and fringe benefits for other employee groups in order to preserve internal equity. Thus, careful control of compensation costs tends to be given lower priority than it will receive at later stages in the life cycle. In these ways, the interaction of tight labor markets, business strategies, and management values contributes to the emergence of several dimensions of the new human resource management system.

. The gradual but steady growth of demand for white-collar, technical, and managerial talent that has been fueled, in part, by the growth of advanced technology industries since 1960 has contributed to the new system in related ways by encouraging more careful management of *individual* motivation, satisfaction, and career development and aspirations (Kanter, 1977). Responsibility for designing management and personnel systems for these more technical and professional workers was given to personnel specialists with training in the behavioral sciences.

Thus, the gradual transformation of the work force and the growing importance of these skills since 1960 laid part of the foundation for the growth in influence of human resource professionals within the firm (Kochan and Cappelli, 1984). The dominant values of these professionals focused on fostering environments that nurtured individual worker motivation, creativity, commitment to the objectives of the firm, and job satisfaction. Moreover, emphasis on individual worker needs and strategies for motivation made these professionals early supporters of job enlargement and rotation, theories of motivation that stressed individual participation in task-related decision making, and compensation systems that tied rewards to individual performance (Hulin and Blood, 1968; Vroom, 1964; Porter and Lawler, 1968).

Government Policy.   Government policy has played an important role in stimulating, diffusing, and institutionalizing changes in personnel and industrial relations practices at a number of critical stages in the evolution of U.S. industrial relations. The role of government can be summarized in three propositions. First, the passage and initial vigorous enforcement of new government policies contribute to the development of new human resource policies and practices. Second, the requirements of these regulations strengthen the role of human resource professionals and planning processes within the firm. Third, declining enforcement pressures by government agencies weaken the ability of human resource professionals to use government policy requirements to maintain influence in management decision making.

The passage of the National Labor Relations Act and the role of the War Labor Board during World War II were critical to the development and institutionalization of collective bargaining (Kochan, McKersie, and Katz, 1985; Jacoby, 1985; Baron et al., 1985). At other times, the absence of an active governmental role has limited diffusion or has contributed to the erosion of employer and/or union commitment to new or emerging policies. For example, the end of the War Labor Board's efforts to promote labor-management production committees after World War II led to the decline of the majority of these committees (Dale, 1949). The withdrawal of the financial and consultant resources provided by the National Commission on Productivity and Quality of Work Life in the mid-1970s also led to the atrophy of most of the work place experiments promoted by this organization (Goodman, 1980).

Similarly, the enactment and the changing patterns of enforcement of various employment standards policies after 1960 have served to increase the priority given to these issues, have strengthened the power of the human resource staff assigned responsibility for managing organizational adaptation to the policy requirements, and have stimulated substantive changes in various recruitment, promotion, and other personnel practices. A Conference Board study concluded, for example, that the passage of equal employment opportunity, safety and health, pension, and other employment regulations served as the most important catalyst for strengthening and professionalizing the emerging human resource management function (Janger, 1977). The government reporting and compliance requirements contained in these laws also led to the collection of demographic data and stimulated their use for human resource forecasting (Milkovich and Mahoney, 1978).

The passage of new public policies, along with the scarcity of technical and managerial talent during the 1960s and part of the 1970s, forced human resource specialists to give high priority to the pressures arising from these environmental forces. This, in turn, increased the leverage of these staff specialists in managerial decision making (Kochan and Cappelli, 1984). By relying on the government requirements, human resource professionals could get top executives and line managers to commit resources and support for policy changes that otherwise would not have occurred because their actions would not have appeared to be necessary for carrying out business strategies and objectives. Thus, government pressure was most influential as an innovative force in the 1960s because the changes it induced proved to be instrumental in improving organizational performance, given the tight labor markets and growing business opportunities for industries employing large numbers of technical and professional employees.

But just as government pressure increased management commitment to these policies and encouraged centralization of responsibilities for adjusting to them in the hands of human resource professionals, the declining commitment and enforcement activity of the Reagan administration in the 1980s eased the pressures on firms and set the stage for personnel and human resource professionals and policies to turn inward to focus on the values and strategic objectives of their top executives.

Union Pressures.   The threat of unionization, and the presence of unions in an organization have served, historically and currently, as important determinants of innovation in personnel policies at all three levels of the framework presented in table 10–1. The effects of unions on management policies have been well documented elsewhere (Slichter, 1941; Slichter, Healy, and Livernash, 1960; Freedman, 1979; Foulkes, 1980; Kochan, 1980; Freeman and Medoff, 1984; Kochan, McKersie, and Katz, 1985); therefore, we do not need to repeat the evidence here. What is most relevant here is an understanding of how the threat effect and the actual effects of unions vary under different competitive conditions or over different stages of a business life cycle. Our central proposition here is that both the threat of unionization and the presence of a union leads to the greatest human resource innovations during growth stages of business life cycles.

In early growth stages of a business, high priority is attached not only to satisfying *individual* worker needs in order to reduce turnover but also to achieving labor peace in order to avoid work stoppages, which impede the firm's ability to exploit its market opportunities or allow competitors to gain market share. For the nonunion firm, the threat of unionization increases the risk of a disruption. Because costs are not so tightly controlled or are not given high priority at this stage, meeting or in some cases exceeding union wage and benefit standards serves as an efficient union avoidance strategy. For the unionized firm, the costs of work stoppages are high relative to the costs of meeting union demands for improved wages and benefits; therefore, unions achieve significant gains through collective bargaining. As businesses mature, however, increased price competition and concern over manufacturing costs lead unionized firms to seek more aggressively to control labor costs. For nonunion firms, the need to cut labor costs is weighed against the priority attached to union avoidance and the severity of the threat of being organized.

Management Values

The values and ideologies of key executives play important roles in filtering pressures that arise from the environment or from business strategies by giving prominence to those policy options that are consistent with the dominant values, well matched to the strategic needs of the firm, and effective in coping with environmental pressures. Our central proposition here is that the greater the diffusion of strong top management values in the firm, the more coherent the human resource management system will be. At the extreme, shared values can be used as substitutes for detailed rules or bureaucratic structures and narrow divisions of labor. This is the essence of clan-type organizations (Ouchi, 1980) or strong organizational cultures (Deal and Kennedy, 1982; Peters and Waterman, 1982). Since the origin of these values or norms usually can be traced to the values of the organization's founders or early leaders, their effects tend to be especially strong in younger organizations and in those where the early successes of the business continue to be reinforced by continued growth and positive economic performance (Schein, 1983). Although the effects of values on human resource management could be illustrated in a number of ways, we will focus on the values of business executives toward unions—or, more specifically, toward union avoidance—since this will demonstrate an important set of interactions among two key variables in our model.

It is widely recognized that the dominant value system of U.S. managers leads them to prefer to operate without unions (Bendix, 1956; Brown and Myers, 1957; Kochan, 1980; Foulkes, 1980). In our research on the growth of the nonunion sector of the U.S. economy since 1960, we found (1) that union avoidance strategies dominated managerial policies toward unionization whenever it was feasible for management to do so; (2) that these strategies led to a higher rate of innovation in human resource practices at the work place, which served to reduce the incentives for workers to organize; and (3) that the combination of union avoidance strategies and work place innovations significantly reduced the number of workers organized in firms between the mid-1970s and early 1980s (Kochan, McKersie, and Chalykoff, 1985). Specifically, we found that the only firms that did *not* place a high priority on union avoidance strategies were those that could not easily pursue union avoidance because they were already highly unionized and dependent on maintaining good relations with their current unions. Those firms that did assign a high priority to union avoidance were more likely to establish nonunion grievance procedures, to use autonomous work teams in nonunion establishments, and to use work sharing instead of layoffs with nonunion employees (Chalykoff, 1985). Together, the effects of these union avoidance strategies and work place innovations reduced the likelihood that a new plant would be organized nearly to zero and, over the six-year time span examined (1977–1983), reduced union membership in these firms by approximately 8 percent. Thus, top management values can have an extremely strong effect on human resource or industrial relations practices at the work place and, in turn, can also affect such outcomes as union membership.

Technology.   Uncertainty over the effects of technology on work organization is perhaps one of the most hotly debated human resource management issues of the decade. Innovations in microelectronics are producing new products and services and are affecting the number of jobs, the nature of the skills required to operate the new technology, and the organization of work in both factories and offices. It is neither possible at this point nor, for our purpose, necessary to predict the net effects of these new technologies on the demand for labor or on skill content. What is critical here is to recognize that the design of work systems for new technologies is itself a choice or decision that is influenced by market characteristics, by the values of decision makers, and by the business strategies of the firm (Walton, 1980). Our basic proposition concerning technology is that the more investment there is in new technology, the more organizations will attempt to introduce flexible work organization designs.

Piore and Sabel (1985) demonstrate, for example, that firms facing uncertain markets and employing flexible manufacturing technologies will seek to design flexible work organization systems of the type characterized earlier as part of the new human resource management model. Evidence from both U.S. and European auto firms (Katz, 1985; Katz and Sabel, 1985) is consistent with this hypothesis. Yet it is not clear that new technology or market uncertainty are the sole or even the dominant forces leading to more flexible work organization designs. The motivation to lower labor costs by reducing employment and eliminating various restrictive work rules may be an equally important factor, especially in mature manufacturing industries such as autos. Moreover, fundamental or complete changes in work organization systems are hard to achieve in established work and organizational settings. Thus, the most flexible work organization systems in the United States, and those reinforced by policies at higher levels of the organization that are consistent with the new model, tend to be found in the newest plants, offices, and firms—particularly in the new "greenfield" plants that recruit and select entirely new work forces (McKersie, 1985; Katz, 1985). Whether these attempts are successful depends on the ability to gain the support of the work force and, where present, the unions representing the workers. Whether the flexibility built into the design of these new work systems atrophies as the employment relationships age will depend on the outcome of the dynamics introduced below.

High Technology Firms and Life Cycle Dynamics

Given the model developed in the preceding sections, it is easy to see why many high technology firms tend to exhibit characteristics of the new human resource management system. They are generally younger firms operating in tight labor markets and developing or growing product markets; they employ high proportions of professional and technical employees; and they are often run by their founders and are able to incorporate new technology into new plants and offices. The rapidly changing and uncertain character of product markets creates further incentives to build flexibility and high levels of participation into work organization and management systems. Business strategies depend on research and development and on the ability to get new products to market quickly. Moreover, although few of the eligible high technology workers are unionized, the priority placed on union avoidance is quite high. Since 1980, relaxation of government enforcement pressures has allowed human resource management specialists to turn their attention inward to respond to the calls from their profession to link human resource policies to internal business strategies and to help shape and diffuse the culture espoused by top executives throughout the organization. This role is reinforced by the burst of books, articles, conferences, and consulting sevices that stand ready to assist human resource managers and line executives in measuring, understanding, shaping, or changing their organization's culture.

Thus, the state of all of the environmental and organizational variables in the model presented in figure 10–1 fit together at this stage in the evolution of high technology industries and firms to create the confluence of conditions that, we have hypothesized, supports the development and stability of the new human resource management model. Whether this model will remain stable over time depends on the extent to which the values supporting the development of this system withstand the pressures that will be experienced as markets mature and as price competition becomes a more important strategic concern and the extent to which environmental conditions either continue to reinforce or challenge the features of the new system. More specifically, as markets mature or price competition intensifies for other reasons, the need to reduce costs will intensify pressures to (1) slow the growth in compensation costs; (2) reduce employment, especially for staff, management, and other indirect labor positions, since this is where the bulk of labor costs lie; and (3) reorganize and rationalize production by seeking lower labor cost environments, especially in firms where production technology is relatively routine and volumes are high.

It is neither inevitable nor likely, however, that firms will simply revert to the traditional human resource management/industrial relations system when faced with these pressures. Those firms that have the diffused and institutionalized values that gave rise to the new system may respond to these economic pressures by holding to their values and absorbing the transition costs until a new business strategy can be put in place that can again support the system. The growth of various voluntary severance, early retirement, and other work-force reducing strategies used by companies that are determined to avoid layoffs or involuntary termination represents an effort to fashion such transition strategies (Kochan and Barocci, 1985).

A case in point is one large high technology firm several of our colleagues have been studying that was facing severe cost competition from overseas and domestic competitors in one of its divisions. At the same time, price competition intensified and demand fell because of a downturn in the business cycle. Rather than resorting to layoffs, which would have violated the employment stabilization policies and values of the firm, various attrition, interplant and intraplant transfer, early retirement, and voluntary severance programs were offered to employees. Also, a larger work force than necessary was carried during the downturn in anticipation of capturing a higher market share when the economy picked up again. Moreover, a new business strategy and accelerated investment in new manufacturing technology were expected to be in place then, which would again stabilize employment. In this example, the organization's values not only overcame the conflict with short-term business strategy pressures but also stimulated the search for a new business strategy that allowed the values to survive.

Summary

The analysis presented in this paper suggests that managerial values and business strategies in the context of specific environmental pressures of the past two decades have been the dominant determinants of a set of policies and practices that we believe constitute a new human resource management system. The confluence of these environmental and organizational forces and the fact that the more innovative organizations were in the growth stages of their business life cycle provided the internal consistency between human resource policy and practices at the different levels of the firms needed to make them durable. As the pressures of government regulation recede in the current environment, and as the price competition in high technology markets intensifies, pressures to adjust these policies will build to the point that firms will need to choose between their desire to maintain commitment to the complete package of policies that give the new system its internal coherence and the need to make incremental policy adjustments in order to respond to shifting business strategies. The more incremental adjustments that are made, the greater the potential there is for internal contradictions among the policies to develop and the greater the potential for conflict among organizational participants who are affected by these policy changes.

The layoffs, wage freezes, temporary shutdowns, and permanent job losses that have occurred in many high technology firms as they adapt to a temporary economic downturn in demand for computer-related products and the intensifying price competition of maturing product markets illustrate the types of pressures for policy adaptation that can lead to the internal contradictions envisioned in our model. Older high technology firms that were viewed as human resource management innovators in the 1960s and 1970s—such as Polaroid, Eastman Kodak, Xerox, and IBM—also have experienced many of the pressures to adapt to more highly competitive product markets and to refocus their business strategies with the objective of lowering manufacturing costs. The strategic adaptations these firms have made in recent years posed challenges to the staffing, compensation, employment stabilization, and work organization practices they had followed during the rapid growth phases of their businesses. Moreover, each of these firms faces the challenge of managing some business units with attractive growth opportunities and, at the same time, having to cope with high-cost operations in other business units that are in more mature and price-competitive markets. Thus, these firms and others that operate in multiple market environments will constantly be facing the decision whether to adapt policies to specific business strategy considerations of different units or to maintain an overall set of policies that are consistent with the values and preferences of those who promoted the initial policy innovations and strategies that produced the new model.

The current passive role of the government and the declining threat posed by unions have created a vacuum that has allowed management to gain the initiative. At other periods of history, government or labor unions have served as the dominant forces for innovation. It is too early to tell where the initiative will lie in producing the next round of human resource innovations, but an answer may be found by observing how the various parties or actors in our industrial relations system respond to the internal contradictions that characterize employment relationships moving from one stage of a business life cycle to another.

Part V
Summary and Conclusions

11
Perspectives and Implications

Archie Kleingartner
Carolyn S. Anderson

The objective of this book has been to examine a number of leading human resource management issues that are shaping the development and functioning of high technology industry. Human resource management in high tech firms is a relatively new area of academic research. There are still substantial gaps in empirical data on human resource systems in high technology, on the extent to which they differ among firms, and on the ways in which they are managed. Thus, this book raises as many questions as it answers. We should be surprised if it were otherwise. The call for further research usually found when a new area of study is explored seems especially apt in this case.

There is no question that high technology based industries are central to the economic future of the United States. At the core of the work force responsible for high technology development are the engineers, scientists, and other professionals who develop the innovations that drive the expansion of high tech. It is not surprising that many of the chapters in this book focus on innovation in human resource policies as they relate to the professional worker component of the work force. The book is also concerned with the elements that shape the human resource requirements, policies, and practices of firms and how they are affected by the fact that high tech firms function in a dynamic environment.

There is a context and rationale to innovations in human resource management. Such innovations do not just happen; there are circumstances under which they are more likely to be established and maintained. Several of our authors see evidence that innovative human resource management systems have become institutionalized in many high tech organizations. For example, in chapter 10, Kochan and Chalykoff suggest that without necessarily having engaged in much conscious planning, many of the newer high tech firms possess the necessary values, business strategies, and environmental conditions that allow innovative human resource practices to flourish. The difficult test for these firms comes, as the case study by Fred Foulkes illustrates, when internal and external pressures (price competition, employment growth, development of new markets and products) upset the confluence of forces associated with the successful entrepreneurial stages that have given rise to innovation. Of course, Kochan and Chalykoff are not suggesting that all or even most newer firms follow innovative human resources practices, but rather that for many of these firms, the basic forces that stimulate and maintain innovative human resource policies and practices are present.

Compared with traditional manufacturing industry, high tech industry tilts in the direction of what Miljus and Smith in chapter 7 call an "adaptive system" and what Kochan and Chalykoff call a "new system," as contrasted with a "traditional system." In neither chapter are the authors simply referring to the high tech firm touted in the popular media as an organization that is growing rapidly and has high profit margins, little product market competition, and little need for unionism and collective bargaining. Rather, they suggest frameworks for examining systems of human resource management practices that apply to a broad range of firms.

James A. Parrott, a thoughtful critic of human resource practices in high tech organizations, agrees with Kochan and Chalykoff's cautions about equating human resource innovation with high technology firms:

> It should also be made explicit that high tech industry is composed of diverse types of employers. Not all high tech firms have the resources or the inclination to carry out innovative employment policies like those developed at IBM, Digital, or Hewlett-Packard. Many smaller high tech firms—particularly those producing commodity-type items, often under contract to the larger end equipment manufacturers—pursue relatively traditional approaches.

This concluding chapter will review the major themes covered in chapters 1 through 10 and will incorporate comments and ideas from a select group of experts (see the list of commentators in the contributors section, below). These commentaries, insightful in their own right, also extend the analysis contained in the chapters, and they bring different perspectives to the issues. In most cases, these perspectives represent the experiences of persons who deal with the issues on a daily basis. The subject matter of the chapter will be covered under the following main headings: the high tech work force, the management of human resources, and industrial relations implications. There is also a concluding comment.

The High Tech Work Force

Belous (chapter 2) and Solmon and La Porte (chapter 3) confirm what other studies have demonstrated—that although high technology industry imposes demands on the labor market that are different from those of other industries, there is little evidence to support the notion of a general shortage of scientists and engineers, now or in the foreseeable future. The United States does not need a new large-scale training program to meet the demands of high tech industry, although there are shortages in selected skill areas, as Belous points out. But the quality of this work force—its ability to keep the nation's high tech industry competitive in the face of international competition—is at issue. In addition, a significant portion of the increase in labor demand will be at the lower level of the skill distribution, not at the graduate level.

Professor Larry Kimbell, director of the UCLA Business Forecasting Project, helps place the Belous forecast in perspective by questioning several key assumptions that seem to underlie most forecasting activity. Much of the engineering of products used in U.S. high technology industries is accomplished with talent that is located abroad. For example, one cost breakdown of the IBM Personal Computer indicated that most of the cost of the materials is incurred overseas. Kimbell notes:

> The total value of its Japanese components, including its printer, exceeds the value of the U.S. components. What we do is box them. They make them, and we box them. We advertise them, we sell them, and we distribute them. Is that the new division of talent and resources? We are going to have to face that issue. How much substitution is there among types of labor?

With respect to the labor requirements for production of high tech products, the industry tends to go where the costs are lowest, other factors being equal. At the same time, high tech industry has tended to expand where there has been a readily available pool of trained scientific and technical workers. These factors have significant implications for the training of high tech employees. As a matter of national policy, there seems little doubt that the commitment is to ensuring that the United States produces the full spectrum of engineering and scientific talent to remain in the forefront of technological change. In regard to how and where the workers to support high tech development are trained, Kimbell asks the following question:

> What should a small state, or even a large state like California, do about training of engineers and scientists? Many places around the country have adopted the policy that they can lure jobs away from California—and not only jobs but the California-trained workers who are to fill them. Every town in the United States has the feeling that it can compete for high technology jobs. That is a zero sum game. Obviously, the companies that have plans to locate and are mobile are going to encourage competition among localities. Similarly, Silicon Valley was an agglomeration point for high tech. We have seen that enormous energy dispersed to other agglomeration points around the country—the Dallas–San Antonio nexus, the Boston area, and so forth. California is concerned about this

dispersion. Maybe it pays California to compete intensely and to continue training engineers, but surely the same is not true of every urban area and state.

Solmon and La Porte review a great deal of data on the implications of high technology development for the educational institutions of our society. Although, like Belous, they do not see an absolute shortage of technically trained people for high technology industry in the near future, they are concerned with such matters as the deterioration of test scores and enrollment in science and math at the high school level and the kinds of training provided at the postsecondary levels. A significant question raised by their chapter has to do with the fit between the preemployment training provided by our educational institutions and the rapidly changing requirements of industry. They suggest that although there is a continuing and perhaps urgent need in some respects to improve the academic quality of the education received in schools and colleges, there is no way that this training can fully anticipate the long-term requirements of industry.

Professor Harry F. Silberman of the Graduate School of Education at UCLA, former chairman of the National Commission on Secondary Vocational Education, comments on the implications of the chapters by Belous and Solmon and La Porte for our educational institutions:

> None of the authors were eager to jump on the high tech training bandwagon, and that is to their credit. . . . They had good and bad news. The good news is that we'll have enough scientists and engineers for our high tech industry. The bad news is that our best people are leaving the university for higher-paying jobs in the private sector, a trend that threatens the quality of the education system that must prepare future scientists and engineers.
>
> Most articles on the educational implications of high technology agree that the quality of education and training must be improved, but the methods by which such improvement is to be achieved either are not specified or are based on the simple assumption that more of the same treatment is the best prescription.
>
> There is some agreement that high tech will require more flexible workers. Such descriptors as adaptiveness, ability to learn, entrepreneurship, inventiveness, creativity, and improvisational skill are commonly mentioned. Social skills and attitudes that improve interpersonal harmony and teamwork are also frequently mentioned. The advantage of such skills lies in their generalizability. They can be transferred to a wide variety of unpredictable work settings; they are not job-specific.
>
> Critics of existing education and training efforts, however, seldom look at the flexibility of students or at their motivation and social skills, only at their academic achievement and length of schooling. It is questionable that more academic study will strengthen the reasoning skills of students. Regarding the effect of such academic courses strengthening the mind: this is the old mental

muscle theory, which likens the brain to a physical muscle that is strengthened by strenuous and painful exercise. Over 60 years ago Professor E. L. Thorndike at Columbia University tested that theory and found it wanting.

Critics also assume that the practical and applied portions of the curriculum—sports, the performing arts, and vocational training—have no general education value, another questionable assumption. These programs have important educational value for students because they provide an alternative to the traditional educational process. I believe it is that alternative process—alternative to passive, rote, competitive, individualistic, teacher-talk, student seatwork approach—which affects student attitudes and motivation and which increases their flexibility and enhances their social skills.

Chapters 3, 4, and 7 emphasize the important role of the high tech firms themselves in ensuring a high-quality work force. In fact, corporations—working with educational institutions, labor unions, and specialized training agencies—have long recognized the need for cooperation in preparing the nonprofessional work force for changing job requirements and occupational mobility. High technology provides some new challenges for the maintenance of a highly qualified work force—especially engineers and scientists, the technical elites on which high technology industry depends for product innovation. The rate of technological change, as several chapters in this book note, is so relentless that there is no way for a university degree program to do more than provide the foundational learning for the professional work force.

In a 1985 study (conducted by the editors) of executives from high technology firms in the Silicon Valley, training and development of technical personnel was ranked very high by respondents in terms of its importance to the firm. It ranked ahead of such matters as compensation/benefits and productivity improvement. We also conducted personal interviews at high technology firms in California and Massachusetts in which we explored the measures these firms take to keep their professionals up-to-date and productive. These firms tend to spend a considerable amount of time and money on activities that combine skill development and broader professional development into a single package. Specific training and development activities that seem to have acquired substantial popularity among high tech firms include:

Allowing employees to use work time to take formal course work at institutions of higher learning and reimbursing their expenses for doing so.

Encouraging employees to pursue advanced degrees. Sometimes this activity is done during work time at full pay; sometimes it is not.

Awarding sabbatical leave to key employees so that they can develop new skills or work on projects of interest to the individual and the firm.

Encouraging employees to report research findings at professional meetings by preparing and delivering papers and publishing in professional journals.

Providing employees free time to work on pet projects, including small expense grants for supplies and other costs to support these projects.

Allowing employees to check out tools and instruments to be used during spare time at home.

Encouraging innovativeness by providing a share of royalties to employees who develop patentable and/or licensable inventions.[1]

Recruiting and retaining the scientific and technical professionals in high technology organizations—and keeping them productive—are not discrete activities but can be viewed as part of a comprehensive philosophy about human resources utilization. A continuous interaction between universities and industry appears to be an essential component. Karl Pister explores the possibilities for such interaction. His discussion of the role of the University of California at Berkeley and Stanford University in the development of Silicon Valley provides a dramatic illustration of the interdependence of universities, industry, and government in high technology development. This pioneering relationship has become the model that other centers of high tech development seek to emulate.

Leaving the university for high technology employment need not spell the end of graduates' active involvement in the university. At some point, they are likely to return to the university for refresher courses or to collaborate on research projects with their university-based colleagues. Futhermore, if the next generation of graduates is to be properly trained, many of the Ph.D.'s employed in high tech firms will have to return to the universities to participate in their instructional and research programs. Universities and high tech industry, with governmental encouragement, are actively promoting these new forms of collaboration.

Management of Human Resources

The emphasis on human resource management policies and practices, especially as they relate to professionals, represents a recognition of the strategic importance of knowledge workers to high tech. Knowledge workers, by what they do and the products they create, change and define what is meant by high technology industry. High technology organizations have a large stake in the development of human resource policies and practices that will maximize the productivity of the professionals and other strategically important workers.

The most heralded examples of internally initiated development are the successful start-ups. Foulkes's study of AutoTel is typical of a successful new firm in which product innovation and great growth occurs over a relatively short time period. (Of course, most start-ups never achieve the success of AutoTel; many end up as case studies of business failures or are absorbed by a more successful competitor.) The experiences of AutoTel are representative of the several ways in which various human resource management procedures—or lack of procedures—are called into question as a firm matures. Although these experiences are in many respects unique to AutoTel, they suggest several general issues that affect the ways in which the human resources of high tech firms are managed.

The perception of the successful high technology start-up firm is that it has a flat organization and little supervisory structure, it avoids distinctions between employees and management, it allows ample leeway for professional workers to pursue their own interests, and it emphasizes such incentive schemes as profit sharing, stock ownership, and the like. AutoTel is an example of a start-up that contributed to this lore.

Over an eight-year period, the work force of AutoTel grew over 1,000 percent, from 150 to 1,773 employees. As this occurred, the firm hired a full-time personnel director, new layers of management were developed, the compensation system was formalized, and the firm instituted a formal performance evaluation system. These and similar responses to growth turned out to have a significant effect on the company. The move toward more bureaucratic organization is often viewed as the inevitable consequence of growth. AutoTel attempted in various ways to retain the informality and esprit de corps that characterized the company during its start-up phase, but with growth came the clear need for a more formalized human resources system. Although the AutoTel study does not explore the firm's response to economic difficulty, other start-ups in similar circumstances have responded as firms have traditionally done, by laying off employees and taking other cost-cutting measures.

A perspective on the human resource practices of established high technology companies—firms that have passed through various stages of development and have dealt successfully with the kinds of pressures AutoTel encountered for the first time—is the focus of the chapter by Miljus and Smith. They, too, are concerned with the factors that influence the human resource function within firms and especially with the role of the human resources manager in the development and implementation of a firm's policies and practices.

In their view, the effective human resource professional is one who is fully knowledgeable about the business strategy of the firm—indeed, who has participated in its formulation. Such a person can make a distinctive contribution by helping line management plan for and acquire the necessary staff and develop the particular human resource practices that hold the greatest likelihood of keeping the professionals productive.

On the basis of interviews with human resource managers and other evidence, Miljus and Smith reject the view held in many quarters that human resource professionals lack an appreciation of and interest in the larger business environment in which high technology firms function. To the contrary, they suggest that human resource professionals in high tech are sensitive to the competitive and organizational environment of the firm and that they take this into account in fashioning human resource policies and practices. The issue regarding organization design and development, which the authors define as "the need to work closely with line managers to shape and maintain organizational conditions that support innovation, change, and employees' continued high performance," is viewed as embodying this new approach to human resource management. Of course, much of human resource professionals' time and attention will continue to be occupied by such traditional responsibilities as salaries and benefits administration, human resource planning, recruitment and staffing, performance evaluation, employee job rights, and the like. Still, working directly with line managers to create and maintain organizational conditions that facilitate innovation and high performance may well represent a significant redefinition in high technology firms of the traditional conception of the responsibilities of the human resources manager.

The chapter by George Milkovich examines the literature on prevalent compensation practices in high technology firms. His analysis provides high technology firms with a basis for determining how their practices compare with innovative compensation practices in the industry. The present state of research into compensation practices in high tech firms has not yet made it possible to prescribe a framework to help managers or human resource professionals select a particular mix of compensation practices best suited for a firm's particular stage of development. However, the data compiled by Milkovich do suggest that pay practices of many high tech firms differ in important respects from those of traditional industries. Such firms are more likely to offer stock ownership plans and are more likely to emphasize special incentives and rewards for key contributors. Still, use of these approaches seems to rest more on faith that desired behavior will result than on a demonstrated impact on performance.

Industrial Relations Implications

It is appropriate that a book on human resource management in high technology firms should stress issues affecting the professional component of the work force. And it is not surprising that the innovations in human resource systems of high tech firms are developed principally to respond to the characteristics, values, and needs of professional workers and their work.

We pointed out in chapter 1 that the realities of much high tech work are frequently at variance with the agenda that busy executives and human resource managers emphasize; issues such as unionism and collective bargaining, grievance procedures, job security, low wages, promotional opportunities, technological change, and layoffs tend to get pushed into the background. It may be that many of high tech's more highly skilled and highly paid workers have been immune from the harsher realities of employment; but for many professionals and much of the nonprofessional work force, high tech falls far short of being problem-free.

Grievance Procedures

David Lewin examines the operation of formal grievance procedures in non-union firms. Comprehensive data are not available on the extent of such procedures in high technology firms, but we do know that a significant number of such firms have developed formal procedures. Most of these firms are looking for ways to deal systematically with grievances in the absence of unionism—and perhaps for ways to help forestall union organization. The implication of Lewin's analysis is that although such procedures can and do work in many instances, they may have significant shortcomings as well. In regard to Lewin's findings, James Parrott comments on the extent to which such procedures provide the guarantees found in the procedures contained in collective bargaining agreements.

> Lewin's analysis of nonunion appeal systems in three large firms generates a number of findings that call into question the extent to which such systems fairly serve workers as an alternative to union grievance procedures. Lewin considers such aspects as the reasons given by nonusers for not availing themselves of the appeal system and examines the performance assessment and career advancement consequences for appeal systems users.
>
> Two salient differences between union grievance systems and the appeal systems utilized in nonunion settings are that in the latter, (1) the worker is on his or her own, without the assistance of an experienced advocate; and (2) since there is no recourse to an outside arbitrator's decision, the company exercises unilateral authority.
>
> Conflict resolution under the nonunion appeals system is strictly on company terms.

Parrott goes on to say:

> In only one of the appeal systems is there provision for employee representation and a role for an outside arbitrator (for such issues as discharge and demotion). Here Lewin notes that "the use of employee representation and arbitration may well stem from the fact that grievance arbitration exists for unionized employees of this firm."
>
> Lewin's examination of the subsequent performance and organizational rewards experienced by appeal system users in all three firms indicates that worker apprehension is, unfortunately, well founded. Although Lewin's methodology does not establish a cause-and-effect relationship, it does show that, at

statistically significant levels, compared with nonfilers, appeal filers had higher voluntary and involuntary turnover rates and were less likely to be promoted in the following year. Moreover, examination of the outcomes of appeal decisions indicated that employees whose appeals were decided in their favor had, in the year following their appeals, significantly lower performance ratings, lower promotion rates, and higher involuntary turnover rates than employees who lost their appeals. These findings were statistically significant for each of the three firms involved.

Parrott concludes by commenting on other related approaches to complaint resolution:

> These strong findings concerning nonunion appeal systems are consistent with the characterization (by a rank-and-file group of IBM workers) of nonunion IBM's "open-door" appeal system as the "open door to the street." The inescapable conclusion seems to be that the appeal system in a nonunion setting can be used effectively by management to punish appeal filers, with punishment meted out in direct relation to the appellant's degree of success. If, in so doing, as Lewin states, management's interest is to "ward off, contain, or displace employee unionization," the modus operandi is not through a mechanism for "conflict resolution" that makes unions unnecessary but through the unilateral exercise of authority and the maintenance of a climate of subservience and fear on the part of workers.

Overall, the comments by Parrott question whether a formal grievance mechanism, unilaterally established by management, can ever provide the same degree of protection for employees that is provided by the grievance procedures regularly negotiated as part of a collective bargaining agreement.

Union Organization in High Tech

Even if we use the broadest definition of high technology industry, suggested by Belous, the extent of union organization in high technology industry is slight. Yet there is evidence from the chapters in this book and from other studies that many of the objective conditions that have historically made unionism attractive— such as low wage rates for many workers and layoffs when firms encounter economic difficulties—are present in high tech firms to a greater degree than is generally assumed. However, the popular perception of high tech firms as good places to work that offer abundant employment and career opportunities has discouraged many unions from even initiating organizing efforts. Parrott speaks from firsthand knowledge on this topic:

> Kassalow is on target in suggesting that extensive media "hype," glorifying conditions in high tech, has made organizing more difficult. My experience has

been that it too often has discouraged unions from taking the industry on. Although I don't think the average high tech production worker is deceived by the "mystique" enveloping high tech—they know they're not sharing in the largesse bestowed on a few—it seems the competition among states and communities to attract high tech development has precluded a more critical look by the press at conditions faced by high tech workers.

The absence of significant union organization may also be attributed, in part, to the inability of unions to devise an organizing strategy that takes into account the particular circumstances of high tech. Evelyn Hunt, a former attorney with the NLRB and a close observer of employee relations in high tech firms, comments on the problems inherent in organizing workers in the Silicon Valley:

> I looked into what was going on in Silicon Valley from the organizing perspective. Apparently, organizing in this industry has been equally unsuccessful in Massachusetts and the Silicon Valley.
>
> The new immigrants—Filipino, Vietnamese, Mexican—have been especially important in the Silicon Valley. The new immigration has an interesting aspect—namely, that workers are not firmly settled. There are no real institutions that union organizers can use to get into the community. There are churches, but there are few neighborhood cultural associations. . . . Up to this time, the union organizers have not been indigenous members of the community. They freely admit that they don't know the culture; they don't know whether it is appropriate to do house visits; they don't know how to get to the people.
>
> I think that in the future, we will see more use made of the indigenous population as organizers. I don't think that there is going to be the traditional plant-gate leaflet type of organizing. That hasn't worked very well.
>
> High tech firms in the Silicon Valley are small entities. Even the large companies are divided up into small groupings of twenty or thirty people in one community. Furthermore, the demographics and traffic patterns create problems. Government urges companies to comply with flextime and different starting and leaving hours. . . . It means that you don't get droves of people leaving all at once. You don't get people going to the bar after the work day, discussing their beefs.
>
> I don't think that union organizing in the Silicon Valley is dead. I think it is dormant. I don't know of any organizing campaigns that are going on there at this time. I do have the impression, from the people I have spoken to, that if it arises again, it is going to take a considerably different form. . . . Whatever form it does take, it is going to be community-oriented, not specifically work place–oriented. In other words, I think that there is going to be an attempt to generate interest around issues—whether it is housing, medical costs, child care—that are not specifically work place related but that will generate interest, and that interest can be used to move along organization in the work place. I think there is going to be, if anything, more attention given to small companies rather than to large companies.

Daniel J. B. Mitchell offers still another slant on the potential for unionization in high tech:

> If unionization ever comes to high tech, it may not come through traditional organizing at all. It may be that the quality circles and the participative management groups and all those kinds of things that have been set up may, in fact, turn out to be the core of something very different, perhaps, from what management had intended.

There is substantial historical precedent for unions to evolve from company organizations initially established by management. The Communication Workers of America, which has recently made a major commitment of funds and staff to organize in high tech firms, is itself an example of an organization that had its start as a company union.

Kassalow's main general conclusion about the future of union organization is that although unions are likely to make inroads, especially among employees at the lower end of the occupational structure, progress will be slow and the character of the bargaining may be quite different from what has been the dominant pattern in the past. With respect to professional and technical workers, unions don't seem any closer to organizing them in high tech firms than in most other private sector industries, where the total proportion organized remains very small despite repeated attempts over the past twenty-five years. Kassalow suggests that the strategies outlined in the 1985 AFL-CIO report on the future of work may offer some new opportunities.

Unions as Change Agents

Union interest in high tech is not confined to matters of organization and representation. The interest of business organizations, government, and educational institutions in high technology development is taken for granted. The unions' stake in this development is frequently overlooked.

Unions are broadly focused, not narrowly focused, institutions. Wherever they occur, plant shutdowns, downgrading of jobs, creation of a two-tiered work force, the impact of technological change, and protection against unfair dismissal are all matters of significance and concern to unions. For example, Kassalow mentions AFL-CIO concerns over the existence of a two-tier work force in high tech, with highly trained and well-paid engineers and technicians on one level and low-wage, low to moderately skilled production and clerical workers on the other. Even if high tech were to be organized, there would be no making up in high tech for the loss of solid middle-income jobs for blue-collar workers in rust-belt industries.

Kassalow sees substantial evidence of a new determination on the part of unions to come to grips with the impact of changes in technology and the employment structure on workers, including those in high tech industries. It is ironic that effective union pressure in some of these areas may actually make the job of organizing high tech workers harder.

Lawrence Littrell of the Northrup Corporation observes a contradiction between the general social and economic agenda of trade unions and their desire to increase membership:

> Any store that sells its products out the front door and gives them away at the side door will sooner or later find few customers coming to the front door. It seems to me that over the years, this is exactly what the labor movement has been doing. In supporting—vigorously supporting—legislation that provides the benefits available, or potentially available, under union contracts to unrepresented employees, they are, in effect, giving away the store. Certainly their members may benefit along with everyone else, but each time such legislation is passed, the need for their service shrinks—however slightly. Over the years, the cumulative effect has been sufficient to raise the question in the mind of the prospective member: "Don't I already have that protection, without having to pay dues?"

Elinor Glenn, a long-time official of the Service Employees International Union, following up on Mr. Littrell's comment, observes:

> Today, I heard something startling. I hear a criticism of the labor movement and a description that says it will not be successful because it not only tries to do things to take care of its own members, it gives away free to other workers things that they ought to pay for by joining the unions—and this is not a good policy for labor. Labor has always fought to improve life for all workers. We care about child care. We care about free education. We care about the forty acres and the mule. We will always care about the people beyond our membership.

Unions emphasize that they will move forward with their traditional mission to achieve progress on broad social and economic fronts, intended to benefit all working people, and to gain recognition to represent workers in the firms where they work. Only time will tell how this scenario will play out in the high tech industries.

A Concluding Comment

We began this chapter with reference to the analytic scheme outlined by Kochan and Chalykoff. It also provides a useful reference for our final comments. All firms, in high technology as in other industries, have human resource management systems, but these systems may differ greatly in their details and in the extent to which they approach what Kochan and Chalykoff call a "new human resources system."

The thrust of their argument is that for the human resources system of a firm to be properly understood, it must be considered at three levels: the executive or strategic level, the middle or functional level, and the work place level. Furthermore, the human resources system of a firm is always embedded in and subject to environmental influences, such as the labor market and the demographics of the work force, government policies, unionism or the threat of unionism, and the nature of technology. Out of the ever-changing environment in which high tech firms function and the confluence of forces that operate at the different levels within firms, are generated the human resources styles, policies, and practices of individual firms.

Kochan and Chalykoff provide illustrative evidence that the emergence of high technology industries is associated with pressures that are redefining many traditional human resource concepts and practices both inside and outside high tech. This point reinforces the claims of several authors that human resource developments in high technology have ripple effects in many other sectors of economic activity. In addition, Kochan and Chalykoff suggest that high technology firms as a group are more apt to be characterized by features of the new human resource management system than firms in other industries are.

The model suggested by Kochan and Chalykoff has obvious attractions for academic researchers who are interested in establishing causal relationships and in understanding the interconnections among various components of a human resources system. However, we believe that their model has practical utility for high technology firms as well. It allows firms to relate features of the model to the development of their own human resources policies and practices and to make adaptations as changes occur in the environment or in the firm, so that congruency across levels is maintained. A congruent human resources system is one in which policies and practices at the three basic levels (strategic, functional, and work place) are complementary to each other and one that gauges accurately the relevant environmental factors. For example, to maintain congruency, the departure of a firm's founder may necessitate a substantial adjustment of the approaches used to maintain the high morale and motivation of the firm's professional workers. Or a nonunion firm that establishes a formal grievance procedure for its employees but then denies an earned promotion to an employee for using it will properly be seen as having been insincere in its motivation for establishing the procedure in the first place.

The chapters in this book have explored many of the key issues relevant to understanding the forces that shape the development of human resource systems in high tech industries. How individual firms respond to these forces will vary. When compared with traditional firms, there is strong evidence that high tech firms are more committed to innovation and experimentation in their human resource practices. This innovation and experimentation derives principally from the assumption of high tech firms that their success is heavily dependent on recruiting, retaining, and motivating those workers who are most strategically important to the firm, whose work is most reliably accomplished under conditions of autonomy, shared decision making, and strong individual and group incentives.

The many references in the high tech literature to such concepts as nonhierarchical worker–management communication, creative compensation systems, employee participation in decision making, and the like, though significant features in many high tech firms, are to be understood as concerned mainly with the desire of high tech firms to devise effective approaches to managing the professional part of the work force.

Although some high tech firms extend their more innovative practices to the blue-collar and other less strategically important nonprofessional workers, this appears to be quite uncommon. Nonprofessionals do not seem to loom large in human resources planning, and when times are difficult, they are dealt with in traditional ways. Ironically, even though unions and collective bargaining have made few inroads in high tech, the cavalier approaches adopted by many of these firms toward the nonprofessional may serve to stimulate interest in protective organization on the part of both nonprofessionals and professionals.

Obviously, there is a need for addtional research on all of the topics covered in this book. However, we feel that the material presented here makes an important contribution to an understanding of human resource management innovation in high tech. The book should prove useful as a starting point for additional research and as a guide for high tech firms as they consider the short- and long-term effectiveness of their own human resource strategies and practices.

Note

1. Based on Kleingartner and Mason (1986).

Bibliography

Abernathy, William J.; Clark, Kim B.; and Kantrow, Alan M., *Industrial Renaissance: Producing a Competitive Future For America.* New York: Basic Books, 1983.

Adams, Larry T."Changing Employment Patterns of Organized Workers." *Monthly Labor Review*, February 1985, 25–31.

AFL-CIO. Committee on the Evolution of Work. *The Changing Situation of Workers and Their Unions.* February 1985.

———. *The Future of Work.* August 1983.

———. *Free Trade Union News*, September 1976.

AFSCME (American Federation of State, County and Municipal Employees). *Facing the Future: AFSCME's Approach to Technology.* Washington, D.C., 1983.

Amatos, Christopher A. "Three Columbus Firms Considering Ireland Plants." *Columbus (Ohio) Dispatch*, 24 May 1985, E7.

American Electronics Association. *AEA Status Report on Engineering and Technical Education.* Palo Alto, Calif. November 1984.

———. *Technical Employment Projections: 1981–1983–1985 Report.* Palo Alto, Calif., 1981.

Anagnoson, J. Theodore, and Revlin, Russell. "Part-Time Employment in California." *California Policy Seminar Final Report Number 8.* Berkeley: University of California at Berkeley, Institute of Governmental Studies, 1985.

Aram, John D., and Salipante, Paul F., Jr. "An Evaluation of Organizational Due Process in the Resolution of Employee/Employer Conflict." *Academy of Management Review* (6 April 1981): 197–204.

Ashenfelter, Orley, and Pencavel, John. "American Trade Union Growth, 1900–1960." *Quarterly Journal of Economics* 83(August 1969): 434–48.

AT&T-CWA. *Memorandum of Understanding.* 10 August 1980.

———. *Memorandum of Understanding.* 1983.

Balfour, Alan. "Five Types of Non-union Grievance Systems." *Personnel* 61(March-April 1984): 67–76.

Balkin, D.B., and Gomez-Mejia, L.R. "Compensation Practices in the High Technology Industry." *Personnel Administration*, 1986.

———. *The Relationship Between Short-Term and Long-Term Pay Incentives and Strategies in the High Technology Industry.* Northeastern University, College of Business, Working Paper, 1985.

Barbash, Jack. "The Impact of Technology on Labor-Management Relations." In *Adjusting to Technological Change*, ed. Gerald G. Somers, E.L. Cushman, and N. Weinberg. New York: Harper & Row, 1963.

Barnett, Chris. "Chips Off The Old Block." *PSA*, September 1983, 67–73, 171–172, 175, 178.

Baron, James N.; Dobbin, Frank R.; and Jennings, Devereaux P. "War and Peace: The Evolution of Modern Personnel Administration in U.S. Industry." Unpublished manuscript, Stanford University, Department of Sociology, 1985.

Beckhard, Richard, and Harris, Reuben T. *Organizational Transitions: Managing Complex Change*. Reading, Mass.: Addison-Wesley, 1977.

Beer, Michael. *Organization Change and Development*. Santa Monica, Calif.: Goodyear, 1980.

Belitsky, A. Harvey. *Productivity and Job Security: Retraining to Adapt to Technological Change*. Washington, D.C.: National Center for Productivity and Quality of Work Life, 1977.

Belous, Richard S. *The Computer Revolution and the U.S. Labor Force*. U.S. Library of Congress, Congressional Research Service. Washington, D.C., 1985.

———. *The Employment Effects Caused by Shifting Sales in Various Industries: An Input/Output Analysis*. U.S. Library of Congress, Congressional Research Service. Washington, D.C., 1984a.

———. *U.S. Wages and Unit Labor Costs in a World Economy*. Report No. 84–172 E, U.S. Library of Congress, Congressional Research Service. Washington, D.C., 1984b.

Bendix, Reinhard. *Work and Authority in Industry*. New York: Wiley, 1956.

Berenbeim, Ronald. *Nonunion Complaint Systems: A Corporate Appraisal*. Report No. 770. New York: The Conference Board, 1980.

Berg, Ivar E. *Education and Jobs: The Great Training Robbery*. New York: Praeger, 1970.

Blanchard, Francis. "Technology, Work and Society: Some Pointers From ILO Research." *International Labor Review*, May–June 1984, 267–76.

Bledstein, Burton J. *The Culture of Professionalism*. New York: Norton, 1976.

Bolle, Mary Jane. *Health Effects of Radiation from Video Display Terminals*. U.S. Library of Congress, Congressional Research Service. Washington, D.C., 1985.

Botkin, James W.; Dimancescu, Dan; and Stata, Ray. *Global Stakes: The Future of High Technology in America*. Cambridge, Mass.: Ballinger, 1982.

Bowles, Samuel, and Gintis, Herbert. *Schooling in Capitalist America*. New York: Basic Books, 1976.

Breckenridge, Charlotte. *High Technology: Growth in The United States Regions, 1977–1981*. Report No. 83-223 E, U.S. Library of Congress, Congressional Research Service. Washington, D.C., 1983.

Broderick, Renae. "Pay Planning and Business Strategy: A Question of Fit." In *Proceedings of the Human Resource Planning Society's Research Symposium*. Plenus Books, forthcoming.

Brooks, Tom. "Job Satisfaction: An Elusive Goal." *American Federationist* 79, no. 10(1972): 1–7.

Brown, Douglas V., and Myers, Charles A. "The Changing Industrial Relations Philosophy of American Management." *Proceedings of the Ninth Annual Winter Meeting of the Industrial Relations Research Association*. Madison, Wis.: IRRA, 1957.

Brown, L. David. *Managing Conflict at Organizational Interfaces*. Reading, Mass.: Addison-Wesley, 1983.

Buffa, Elwood S. "Making American Manufacturing Competitive." *California Management Review*, Spring 1984, 29–46.

Bureau of National Affairs. *Daily Labor Report*, 6 May 1985a, A2–A5.

———. *Daily Labor Report*, 8 May 1985b, C1–C3.

———. *Productivity Improvement Programs.* PPF Report No. 138, September 1984a.

———. "UAW-GM Report." *Daily Labor Report*, 27 September 1984b.

———. *VDTs in the Workplace: A Study of Effects on Employment.* Washington, D.C., 27 September 1984c.

———. "Summary of Tentative Agreement between UAW and General Motors Corporation." *Daily Labor Report*, 25 March 1982, E1–E6.

———. *Special Report: Teachers and Labor Relations, 1979–1980.* Washington, D.C., 1980.

Burgan, John U. "Cyclical Behavior of High Tech Industries." *Monthly Labor Review*, May 1985, 9–15.

Burns, T., and Stalker, G.M. *The Management of Innovation.* London: Tavistock Institute, 1961.

Business–Higher Education Forum. *Highlights of the Winter 1984 Meeting.* Washington, D.C.: American Council on Education, 1984.

California Postsecondary Education Commission. *Foreign Graduate Students in Engineering and Computer Science at California's Public Universities.* CPEC Report 83–37. Sacramento, December 1985.

Carnegie Foundation. *Corporate Classrooms, The Learning Business.* Special Report, 1985.

Carnevale, Anthony P., and Goldstein, Harold. *Employee Training: Its Changing Role and Analysis of New Data.* Washington D.C.: ASTD Press, 1983.

Chalykoff, John B. "Industrial Relations at the Strategic Level: Indicators and Outcomes." Unpublished manuscript, MIT, Sloan School of Management, 1985.

Chamberlain, Neil W., and Kuhn, James W. *Collective Bargaining*, 3d ed. New York: McGraw-Hill, forthcoming.

Chamot, Dennis. "Scientists and Unions: The New Reality." Reprinted from *The American Federationist*, September 1974; revised March 1978.

Chamot, Dennis, and Baggett, John M., eds. *Silicon, Satellites and Jobs.* Washington, D.C.: AFL-CIO, 1979.

Committee on Human Resources. "Human Resources and Competitiveness" In *Global Competition: The New Reality*, Vol. 2, The President's Commission on Industrial Competitiveness, January 1985.

"The Computer Era—It's Here and New." In *Silicon Valley U.S.A.*, special reprint issue of the *San Francisco Chronicle*, 27 September 1980.

Congressional Record, 17 April 1985, S431.

Cook, Frederick & Company. *Entrepreneurial Incentives Survey – 24 Major Firms*, May 1985.

Cooke, William N. "Determinants of the Outcomes of Union Certification Elections." *Industrial and Labor Relations Review* 36(April 1983): 402–14.

Cooper, Michael R. "Changing Employee Attitudes: Fast Growth vs. Slow Growth Organizations." In *A Strategic Report: Linking Employee Attitudes and Corporate Culture To Corporate Growth and Profitability.* Philadelphia: Hay Management Consultants, 1984.

Crystal, G.S. *Executive Compensation*. Englewood Cliffs, N.J.: Prentice-Hall, 1984.

Dale, Ernest. *Greater Productivity Through Labor-Management Cooperation*. New York: American Management Association, 1949.

"Dana University: Offering Unique Industrial Training at Dana Corporation." *Training and Development Journal*, March 1977, 46–48.

Dauffenbach, Robert C., and Fiorito, Jack. *Projections of Supply Scientists and Engineers to Meet Defense and Nondefense Requirements, 1981–1987: A Report to the National Science Foundation*. Stillwater: Oklahoma State University, 1983.

Deal, Terrence E., and Kennedy, Allan A. *Corporate Cultures*. Reading, Mass.: Addison-Wesley, 1982.

Deming, W. Edwards. *Quality, Productivity and Competitive Position*. Cambridge: MIT Center for Advanced Engineering Study, 1982.

Derber, M. "Crosscurrents in Workers' Participation." *Industrial Relations* 9(February 1970): 123–36.

Deutch, Shea, and Evans, Inc. *High Technology Recruitment Index: Year End Review and Forecast*. New York, 1983.

Devanna, Mary Anne; Fombrun, Charles; Tichy, Noel; Warren, Lynn; and Warren, E. Kirby. *Human Resources Management: Issues for the 1980s*. New York: Columbia Business School, Center for Research in Career Development, 1983.

Dickens, William T., and Leonard, Jonathan S. "Accounting for the Decline in Union Membership, 1950–1980." *Industrial and Labor Relations Review* 38(April 1985): 323–34.

Dodd, Martin H. *Professional Workers and Unionization: A Data Handbook*. Prepared for the Department for Professional Employees, AFL-CIO, January 1979.

Dorfman, Nancy S. "Massachusetts' High Technology Boom in Perspective: An Investigation of Its Dimensions, Causes and the Role of New Firms." CPA 82–2. Cambridge, Mass.: MIT Center for Policy Alternatives, April 1982, 17–20.

Dornbusch, Sanford M., and Scott, W. Richard. *Evaluation and the Exercise of Authority*. San Francisco: Jossey-Bass, 1977.

Drucker, Peter F. "The Discipline of Innovation." *Harvard Business Review*, May-June 1985, 67–72.

Drucker, Peter. "The 'Re-Industrialization' of America." *Wall Street Journal*, 13 June 1980.

Duchin, Faye, and Szyld, Daniel B. *A Dynamic Input-Output Model with Assured Positive Output*. New York: New York University, Institute for Economic Analysis, 1984.

Dvork, Eldon J. "Will Engineers Unionize?" *Industrial Relations* 2(May 1963): 45–66.

Dyer, Lee. "Studying Human Resource Strategy: An Approach and an Agenda." *Industrial Relations* 23(Spring 1984).

Dyer, Lee; Foltman, Felician; and Milkovich, George. "Contemporary Employment Stabilization Practices." In *Human Resource Management and Industrial Relations*, ed. Thomas A. Kochan and Thomas A. Barocci. Boston: Little, Brown, 1985.

Early, Steve, and Wilson, Rand. "High Tech Workers Form Network." *The Citizen Advocate* (published by Massachusetts Fair Share) 5(Spring 1983).

Education Commission of the States, 1982. *The Information Society: Are High School Graduates Ready?* Denver: Education Commission of the States, 1982.

Eurich, Nell P. *Corporate Classroom: The Learning Business.* Princeton, N.J.: Carnegie Foundation for the Advancement of Teaching, 1985.

Fein, Mitchell. "Motivation for Work." In *Handbook of Work, Organization and Society,* ed. Robert Dubin. Chicago: Rand McNally College Publishing Company, 1976, 465–530.

Fombrun, Charles. "Conversation with Reginald H. Jones and Frank Doyle." *Organizational Dynamics,* Winter 1982, 42–63.

Foulkes, Fred K. *Personnel Policies in Large Nonunion Companies.* Englewood Cliffs, N.J.: Prentice-Hall, 1980.

Freedman, Audrey. *The New Look in Wage Policy and Employee Relations.* Report No. 865. New York: The Conference Board, 1985.

———. *Managing Labor Relations.* New York: The Conference Board, 1979.

Freeman, John; Carroll, Glen R.; and Hannan, Michael T. "Age Dependence in Organizational Death Rates." *American Sociological Review* 48(1983): 692–710.

Freeman, Richard A. "A Cobweb Model of the Supply and Starting Salary of New Engineers." *Industrial and Labor Relations Review,* January 1976, 236–48.

Freeman, Richard B., and Medoff, James L. *What Do Unions Do?* New York: Basic Books, 1984.

French, Wendell L., and Bell, Cecil H. *Organization Development: Behavioral Science Interventions for Organization Improvement.* Englewood Cliffs, N.J.: Prentice-Hall, 1984.

Friedman, Sheldon. "Negotiated Approaches to Job Security." In "Industrial Relations Research Association Proceedings," *Labor Law Journal,* August 1985, 553–57.

Friedson, Eliot. *Professional Dominance: The Social Structure of Medical Care.* New York: Atherton Press, 1970.

Galbraith, Jay R. "Evolution Without Revolution: Sequent Computer Systems." *Human Resource Management* 24(Spring 1985): 9–24.

———. *Designing Complex Organizations.* Reading, Mass.: Addison-Wesley, 1973.

Galbraith, J.R., and Nathanson, D. *Strategy Implementation.* St. Paul, Minn.: West, 1978.

Gandz, Jeffrey. *The Role of the Industrial Relations Manager in Conflict Resolution.* Working Paper, University of Western Ontario School of Business Administration, July 1978a.

———. *Union-Management Relationships, Grievance Rates, and Arbitration.* Working Paper, University of Western Ontario School of Business Administration, July 1978b.

Giroux, Henry A. "Theories of Reproduction and Resistance in the New Sociology of Education: A Critical Analysis." *Harvard Educational Review* 52(August 1983): 257–93.

Glaser, Barney G. *Organizational Scientists: Their Professional Careers.* Indianapolis: Bobbs-Merrill, 1964.

GM-UAW. *Agreement Between General Motors and the UAW.* 21 March 1982.

Goldstein, Bernard. "Some Aspects of the Nature of Unionism Among Salaried Professionals in Industry." *American Sociological Review* 20(1955): 199–205.

Gomberg, William. "Job Satisfaction: Sorting Out the Nonsense." *American Federationist* 80, no. 6(1973): 14–19.

Gomez-Mejia, L.R., and Balkin, D.B. *Managing a High Tech Venture*. Working Paper, University of Florida, Graduate School of Business, 1985.

Goodman, Paul S. "Quality of Work Life Projects in the 1980s." In *Proceedings of the 1980 Spring Meeting of the Industrial Relations Research Association*. Philadelphia: IRRA, 1980.

Gordon, Bruce F., and Ross, Ian C. "Professionals and the Corporation." *Research Management 5*, no. 6(1962): 493–505.

Gorovitz, Elizabeth. "Employee Training: Current Trends, Future Challenges." *Training and Development Journal*, August 1983, 25–29.

Gosh, A., and Sengupta, A.K. *Income Distribution and the Structure of Production*. New Delhi: South Asian Publishers, 1984.

Gouldner, Alvin W. "Cosmopolitans and Locals: Toward an Analysis of Latent Social Roles—I & II." *Administrative Science Quarterly* 2(1957), 281–305, (1958), 444–67.

Gravelle, Jane G. *Effective Tax Burdens of Human Capital Investment Under the Income Tax and Proposed Consumption Tax*. Report No. 84–741 E, U.S. Library of Congress, Congressional Research Service. Washington, D.C., 1984.

Grunwald, Joseph, and Flamm, Kenneth. *The Global Factory: Foreign Assembly in International Trade*. Washington, D.C.: Brookings, 1985.

Hall, Robert E. "The Importance of Lifetime Jobs in the U.S. Economy." *American Economic Review* 72(September 1982): 716–24.

Hay Group. "Survey of Incentive Practices in High Technology Firms." In *Ideals and Trends*. Chicago: Commerce Clearing House, Inc., 1985.

Heneman, Herbert G., III, and Sandver, Marcus H. "Predicting the Outcome of Union Certification Elections: A Review of the Literature." *Industrial and Labor Relations Review*, July 1983, 537–59.

Herbert, Tom, and Coyne, John. *Getting Skilled: A Guide to Private Trade and Technical Schools*. New York: Dutton, 1980.

Herbst, P.G. *Alternatives to Hierarchies*. Leiden, Netherlands: Martinus Nijhoff, 1976.

High Tech Workers Monitor. April-May 1983.

Hill, John. *From Subservience to Strike: Industrial Relations in the Banking Industry*. St. Lucia, Australia: University of Queensland Press, 1982.

Hirschman, Albert O. *Exit, Voice and Loyalty*. Cambridge: Harvard University Press, 1970.

Horovitz, Bruce, and McClenahen, John S. "Restructuring American Industry." *Industry Week*, 1 November 1982, 36–48.

Howard, Robert. "Second Class in Silicon Valley." *Working Papers* 8(September-October 1981): 20–31.

Huddleston, Kenneth, and Fenwick, Dorothy. "The Productivity Challenge: Business Education Partnerships." *Training and Development Journal* 37(April 1983): 96–100.

Hughes, Everett C. *Men and Their Work*. New York: Free Press, 1958.

Hulin, Charles L., and Blood, Milton R. "Job Enlargement, Individual Differences, and Worker Responses." *Psychological Bulletin* 69(1968): 41–55.

Ichniowski, Casey. *Industrial Relations and Economic Performance: Grievances and Productivity*. Working Paper No. 1367, National Bureau of Economic Research, June 1984.

IUE Local 201, Lynn, Massachusetts. *Electrical Union News*, 1 February 1985.
———. *Electrical Union News*, 15 June 1984.
Jacoby, Sanford M. "Progressive Discipline in American Industry: Origins, Development, and Consequences." In *Advances in Industrial and Labor Relations*, Vol. 3, ed. David Lewin and David B. Lipsky. Greenwich, Conn.: JAI Press, forthcoming.
———. *Employing Bureaucracy: Managers, Unions and the Transformation of Work in American Industry, 1900–1945*. New York: Columbia University Press, 1985.
———. "The Origins of Internal Labor Markets in American Manufacturing Firms." Ph.D. dissertation, University of California at Berkeley, 1981.
Janger, Alan. *The Personnel Function*. New York: The Conference Board, 1977.
Kanter, Rosabeth Moss. "Variations in Managerial Career Structures in High Technology Firms: The Impact of Organizational Characteristics on Internal Labor Market Patterns." In *Internal Labor Markets*, ed. Paul Osterman. Cambridge: MIT Press, 1985.
———. *The Change Masters: Innovation for Productivity in the American Corporation*. New York: Simon and Schuster, 1983.
———. *Men and Women of the Corporation*. New York: Basic Books, 1977.
Kanter, Rosabeth Moss, and Buck, John D. "Reorganizing Part of Honeywell: From Strategy to Structure." *Organizational Dynamics*, Winter 1985, 5–25.
Katz, Harry C. *Shifting Gears: Changing Labor Relations in the U.S. Automobile Industry*. Cambridge: MIT Press, 1985.
Katz, Harry C.; Kochan, Thomas A.; and Weber, Mark R. "Assessing the Effects of Industrial Relations and Quality of Working Life Efforts on Organizational Effectiveness." *Academy of Management Journal* 28(September 1985): 509–26.
Katz, Harry C., and Sabel, Charles F. "Industrial Relations and Industrial Adjustment: The World Car Industry." *Industrial Relations* 24, no. 3(1985) : 295–315.
Kelley, Robert E. *The Gold Collar Worker: Harnessing the Brainpower of the New Workforce*. Reading, Mass.: Addison-Wesley, 1985.
Kennedy, Thomas. *Automation Funds and Displaced Workers*. Cambridge: Harvard University Press, 1962.
Kerr, Clark, and Rosow, Jerome M., eds. *Work in America: The Decade Ahead*. New York: Van Nostrand Reinhold, 1979.
Kerr, J.L. "Diversification Strategies and Managerial Rewards: An Empirical Study." *Academy of Management Journal* 28(1985): 155–79.
———. *Reward System Design and Strategy: A Review and Agenda for Research*. Working Paper, Southern Methodist University, School of Business, 1985.
Kleingartner, Archie. "Collective Bargaining between Salaried Professionals and Public Sector Management." *Public Administration Review* 33(March-April 1973): 165–72.
Kleingartner, Archie, and Mason, R. Hal. "Management of Creative Professionals in High Tech Firms." In *Industrial Relations Research Association Proceedings*, April 1986, 508–20.
Koch, Donald L. "Productivity: The Micro Solution." *Economic Review*, Federal Reserve Bank of Atlanta, September 1983, 34–41.
Koch, Donald L.; Cox, William N., Steinhauser, Delores; and Whigham, Pamela V. "High Technology: The Southeast Reaches Out for Growth Industry." *Economic Review*, Federal Reserve Bank of Atlanta, September 1983, 4–19.

Koch, Donald L.; Steinhauser, Delores W.; McCrackin, L.; and Hart, Kathryn. "High-Performance Companies in the Southeast: What Can They Teach Us?" *Economic Review*, Federal Reserve Bank of Atlanta, April 1984, 4–24.

Kochan, Thomas A. *Collective Bargaining and Industrial Relations*. Homewood, Ill.: Irwin, 1980.

Kochan, Thomas A., and Barocci, Thomas A., eds. *Human Resource Management and Industrial Relations*. Boston: Little, Brown, 1985.

Kochan, Thomas A., and Cappelli, Peter. "The Transformation of the Industrial Relations and Personnel Function." In *Internal Labor Markets*, ed. Paul Osterman. Cambridge: MIT Press, 1984.

Kochan, Thomas A., and McKersie, Robert B. "Collective Bargaining—Pressures for Change." *Sloan Management Review* 24(Summer 1983): 59–66.

Kochan, Thomas A.; McKersie, Robert B.; and Cappelli, Peter. "Strategic Choice and Industrial Relations Theory." *Industrial Relations*, Winter 1984: 16–39.

Kochan, Thomas A.; McKersie, Robert B., and Chalykoff, John B. "Corporate Strategy, Workplace Innovations and Union Members: Implications for Industrial Relations Theory and Practice." Unpublished manuscript, MIT, Sloan School of Management, 1985.

Kochan, Thomas A.; McKersie, Robert B., and Katz, Harry C. *Strategic Choices in U.S. Industrial Relations*. New York: Basic Books, 1985.

———. "U.S. Industrial Relations in Transition: A Summary Report." Paper presented at the annual meeting of the Industrial Relations Research Association, Dallas, 1984.

Kohl, George. "Changing Competitive and Technology Environments in Telecommunications." In *Labor and Technology: Union Response to Changing Environments*, ed. Donald Kennedy, Charles Craypo, and Mary Lehman. University Park: Pennsylvania State University, 1982.

Koptis, George. "Factor Prices in Industrial Countries." *International Monetary Fund Staff Papers*, September 1982.

Kornhauser, Arthur; Dubin, Robert; and Ross, Arthur. *Industrial Conflict*. New York: McGraw-Hill, 1954.

Kornhauser, William. *Scientists in Industry*. Berkeley: University of California Press, 1962.

Kuhn, James W. *Bargaining in Grievance Settlement*. New York: Columbia University Press, 1961.

Kureczka, Joan. "Biotechnology and Government Regulation: Overview and Recommendations." *Bulletin of the Institute of Governmental Studies* 25(June 1984): 1–13.

Landen, D.L., and Carlson, Howard C. "Strategies For Diffusing, Evolving, and Institutionalizing Quality of Work Life at General Motors." In *The Innovative Organization*, ed. Robert Zager and Michael P. Rosow. New York: Pergamon Press, 1982.

Lawler, Edward E. "Creating High Involvement Work Organizations." In *Human Resource Productivity in the 1980s*, ed. Eric G. Flamholtz. Los Angeles: UCLA Institute of Industrial Relations, 1982.

———. "The Strategic Design of Reward Systems." In *Handbook of Organization Behavior*, ed. J. Lorsch. Englewood Cliffs, N.J.: Prentice-Hall, forthcoming.

Lawler, John J. "The Influence of Management Consultants on the Outcome of Union

Certification Elections." *Industrial and Labor Relations Review* 38(October 1984): 38–51.

Lawler, John J., and West, Robin. "Attorneys, Consultants, and Union-Avoidance Strategies in Representation Elections." Paper presented at the Second Berkeley Conference on Industrial Relations, Berkeley, February 1985.

Lawrence, Paul R., and Dyer, Davis. *Renewing American Industry*. New York: Free Press, 1983.

Lechner, Herbert D. "What Really Made Silicon Valley? New Study Reveals Keys." *California Business*, February 1985, 76–77.

Leontief, Wassily. "Why Economics Needs Input-Output Analysis." *Challenge*, March-April 1985.

———. "Technological Advance, Economic Growth and the Distribution of Income." *Population and Development Review* 9(September 1983): 403–10.

Leontief, Wassily, and Duchin, Faye. *Automation: The Changing Pattern of United States Exports and Imports, and Their Implications for Employment*. New York: New York University, Institute for Economic Analysis, 1985.

———. *The Impact of Automation on Employment, 1963–2000*. New York: New York University, Institute of Economic Analysis, 1984.

Leontief, Wassily; Duchin, Faye; and Szyld, Daniel B. "New Approaches in Economic Analysis." *Science*, April 1985.

Levin, Henry M. "Jobs: A Changing Workforce, A Changing Education?" *Change* 16(October 1984a): 32–37.

———. "Improving Productivity through Education and Technology." Project Report No. 84–A25. Stanford, Calif.: Institute for Research on Educational Finance and Governance, November 1984b.

Levin, Henry M., and Rumberger, Russell W. "The Educational Implications of High Technology." Project Report No. 83–A4, Stanford, Calif.: Institute for Research on Educational Finance and Governance, February 1983.

Levine, Marsha. *Corporate Education and Training*. Washington, D.C.: American Enterprise Institute for Public Policy Research, 1982.

Levitan, Sar A., and Werneke, Diane. *Productivity: Problems, Prospects, and Policies*. Baltimore: Johns Hopkins University Press, 1984.

Lewicki, Roy J., and Litterer, Joseph A. *Negotiation*. Homewood, Ill.: Irwin, 1985.

Lewin, David, and Feuille, Peter. "Behavioral Research in Industrial Relations." *Industrial and Labor Relations Review 36(April 1983): 341–60*.

Lewin, David, and Peterson, Richard B. *The Modern Grievance Procedure in the American Economy: A Theoretical and Empirical Analysis*. Westport, Conn.: Quorum Books, forthcoming.

Lusterman, Seymour. *Education in Industry*. New York: The Conference Board, 1977.

Lynton, Ernest A. "Improving Cooperation Between Colleges and Corporation." *Educational Record*, Fall 1982, 20–25.

Maidique, Modesto A., and Hayes, Robert H., "The Art of High-Technology Management." *Sloan Management Review* 25, no. 2(1984): 17–31.

Marcson, Simon. *The Scientist in American Industry*. Princeton, N.J.: Princeton University Press, 1960.

Massachusetts High Tech Council. *Why a Technology Council*. Boston, 1984.

Mayo, John S. "The Evolution of Information Technologies." In *Information Technologies and Social Transformation*. Washington, D.C.: National Academy Press, 1985.

McCartney, Laton. "Our Newest High-Tech Export: Jobs." *Datamation*, May 1983, 114–18.

McKersie, Robert B. "Start-Up of a New Plant: Some Procedural Issues." In *Human Resource Management and Industrial Relations*, ed. Thomas A. Kochan and Thomas A. Barocci. Boston: Little, Brown, 1985.

Medoff, James L. "The Structure of Hourly Earnings Among U.S. Private Sector Employees: 1973–1984." Unpublished manuscript, Harvard University, 1984.

Medoff, James L., and Strassman, Paul A. "About the 'Two-Tier' Work Force and Growth of Low-Pay Jobs." Unpublished manuscript, Harvard University, 1985.

Miles, Raymond E., and Snow, Charles C. "Designing Strategic Human Resources Systems." *Organizational Dynamics*, Summer 1984, 36–52.

Milkovich, G.T., and deBejar, G. *Business Strategies, Human Resource Strategies and Performance*. Working Paper, Cornell University, 1985.

Milkovich, G.T., and Newman, J.M. *Compensation*. Plano, Texas: Business Publications, Inc., 1984.

Milkovich, George T., and Mahoney, Thomas A. "Human Resource Planning Models: A Perspective ." *Human Resource Planning* 1(1978): 19–30.

Miller, George A. "Professionals in Bureaucracy: Alienation Among Industrial Scientists and Engineers." *American Sociological Review* 32 (October 1967): 755–68.

Miller, Michael. "Unions Curtail Organizing in High Tech." *Wall Street Journal*, 13 November 1984, 35, 40.

Miller, Ronald E., and Blair, Peter D. *Input/Output Analysis: Foundations and Extensions*. Englewood Cliffs, N.J.: Prentice-Hall, 1985.

Mintzberg, Henry. "Review of Research on Corporate Strategy." Seminar delivered at Sloan School of Management, MIT, April 1985.

Montagna, Paul D. "Professionalization and Bureaucratization in Large Professional Organizations." *American Journal of Scoiology* 74(1968): 138–45.

Moore, Wilbert E. *The Professions: Roles and Rules*. New York: Russell Sage Foundation, 1970.

Moore, William J., and Newman, Robert. "On the Prospects for American Trade Union Growth: A Cross-Section Analysis." *Review of Economics and Statistics* 57(November 1975): 435–45.

Murphy, Kevin. *Technological Change Clauses in Collective Bargaining Agreements*. Washington, D.C.: AFL-CIO, 1981.

National Academy of Sciences. *International Competition in Advanced Technology: Decisions for America*. Washington, D.C.: National Academy Press, 1983a.

———. *Personnel Needs and Training for Biomedical and Behavioral Research*. Washington, D.C.: National Academy Press, 1983b.

National Research Council. *Engineering Education and Practice in the United States*. Washington, D.C.: National Academy Press, 1985.

———. *Doctorate Recipients from United States Universities: Summary Report 1983*. Washington, D.C.: National Academy Press, 1983.

National Science Board, National Science Foundation. *Science Indicators 1982.* Washington, D.C.: U.S. Government Printing Office, 1983.

National Science Foundation. *Characteristics of Doctoral Scientists and Engineers in the United States: 1982, 1983, 1985.* Washington, D.C.: U.S. Government Printing Office.

———. *Projected Response to the Science, Engineering, and Technical Labor Market to Defense and Nondefense Needs: 1982–87.* Washington, D.C.: U.S. Government Printing Office, 1984.

———. *Federal Funds for Research and Development: Fiscal Years 1982, 1983, and 1984. Washington, D.C.: Division of Science Resource Studies, 1983.*

———. *Foreign Participation in U.S. Science and Engineering Higher Education and Labor Markets.* Washington, D.C.: U.S. Government Printing Office, 1981.

———. *Characteristics of Recent Science/Engineering Graduates: 1980.* NSF 82-313. Washington, D.C., 1982, 14, 30, and 46–62.

Normin, Colin. *The God that Limps: Science and Technology in the Eighties.* New York: W.W. Norton, 1981.

"Norway—New Technology and Data." *European Industrial Relations Review*, no. 106(November 1982): 12–15.

Nulty, Leslie E. "Case Studies of IAM Local Experiences with the Introduction of New Technologies." In *Labor and Technology: Union Response to Changing Environments*, ed. Donald Kennedy, Charles Craypo, and Mary Lehman. University Park: Pennsylvania State University, 1982.

Oakes, Jeannie. "Classroom Social Relationships: Exploring the Bowles and Gintis Hypothesis." *Sociology of Education* 55, no. 4(1982), 197–211.

Odiorne, George S. *Strategic Management of Human Resources.* San Francisco: Jossey-Bass, 1984.

Office of Technology Assessment. *Computerized Manufacturing Automation, Employment, Education and the Workplace.* Washington, D.C.: U.S. Government Printing Office, 1984.

———. *Automation and the Workplace: Selected Labor Education and Training Issues.* Washington, D.C.: U.S. Government Printing Office, 1983.

O'Toole, James, ed. *Work and the Quality of Life.* Cambridge: MIT Press, 1974.

Ouchi, William. *Theory Z: How American Business Can Meet the Japanese Challenge.* Reading, Mass.: Addison-Wesley, 1981.

———. "Markets, Bureaucracy and Clans," *Administrative Science Quarterly* 25(1980): 129–41.

Pacific Studies Center. *Silicon Valley: Paradise or Paradox?* Mountain View, Calif., 1977.

Parrott, James. "High Technology Electronics in Massachusetts: Its Development and Outlook." 9 May 1981, mimeographed.

Peach, David A., and Livernash, Robert E. *Grievance Initiation and Resolution.* Cambridge: Harvard University Press, 1974.

Peters, Thomas J. "Strategy Follows Structure: Developing Distinctive Skills." *California Management Review*, Spring 1984, 111–25.

Peters, Thomas J., and Waterman, Robert H., Jr. *In Search of Excellence.* New York: Harper & Row, 1982.

Piore, Michael J., and Sabel, Charles F. *The Second Industrial Divide*. New York: Basic Books, 1985.

Ponak, Allen M. "Unionized Professionals and the Scope of Bargaining." *Industrial and Labor Relations Review* 34(1981): 396–407.

Porter, Lyman W., and Lawler, Edward E. *Managerial Attitudes and Performance*. Homewood, Ill.: Irwin, 1968.

Porter, Michael E. *Competitive Strategy: Techniques for Analyzing Industries and Competitors*. New York: Free Press, 1980.

Portes, Alejandro. "Industrial Development and Labor Absorption: A Reinterpretation." Presentation at UCLA Latin American Studies Center, May 1985.

Public Agenda Foundation. "American Management Style Undermines Work Ethic." News release, 5 September 1983.

Pyle, William. "Accounting for People." In *Managing Advancing Technology: Creating An Action Team in R&D*, Vol. 2. New York: AMA, 1972.

Quinn, James Brian. "Managing Innovation: Controlled Chaos." *Harvard Business Review* May-June 1985, 73–84.

Ramo, Simon. "U.S. Technology Policy—An Engineer's View." *The Bridge* 14(Fall 1984): 2–8.

Rappaport, A. "Executive Incentive vs. Corporate Growth." *Harvard Business Review* 59(1978): 81–88.

Riche, Richard W.; Hecker, Daniel E.; and Burgan, John U. "High Technology Today and Tomorrow: A Small Slice of the Employment Pie." *Monthly Labor Review*, November 1983, 50–58.

Riggs, Henry E. *Managing High Technology Companies*. Belmont, Calif.: Lifetime Learning, 1983.

"Right to Know: Is it Your 'Right'?" *Boston Globe*, 7 June 1983, 53, 55.

Rogers, Everett M., and Larsen, Judith K. *Silicon Valley Fever: Growth of High Technology Culture*. New York: Basic Books, 1984.

Rowe, Mary P., and Baker, Michael. "Are You Hearing Enough Employee Concerns?" *Harvard Business Review*, May-June 1984, 127–35.

Rozen, Miriam. "Wanted: High Tech Engineers." *Dun's Business Month*, March 1985, 35–36.

Rubin, J.Z., and Brown, B.R. *The Social Psychology of Bargaining and Negotiation*. New York: Academic Press, 1975.

Rumberger, Russell W., and Levin, Henry M. "Forecasting the Impact of New Technologies on the Future of Job Markets." Project Report No. 84–A4. Stanford, Calif.: Institute for Research on Educational Finance and Governance, February 1984.

Salter, M.A. "Tailor Incentive Compensation to Strategy." *Harvard Business Review* 51(1973): 94–102.

Sanders, W. Jerry. Speech given at dedication of CAD/CAM Facility, University of California, Berkeley, September 1985.

Sayles, Leonard R. *The Behavior of Industrial Work Groups*. New York: Wiley, 1958.

Schein, Edgar. "The Role of the Founder in Creating Organizational Culture." *Organizational Dynamics*, Summer 1983, 13–28.

Schuster, Jay. *Management Compensation in High Technology Companies*. Lexington, Mass.: Lexington Books, 1984.

Scott, Allen. *High Technology Industry and Territorial Development: The Rise of the Orange County Complex.* Working Paper Series 85. Los Angeles: UCLA Institute of Industrial Relations, 1985a.

———. *The Semi-Conductor Industry in South-East Asia: Organization, Location and the International Division of Labor.* Working Paper Series 101. Los Angeles: UCLA Institute of Industrial Relations, 1985b.

Scott, W. Richard. *Organizations—Rational, Natural, and Open Systems.* Englewood Cliffs, N.J.: Prentice-Hall, 1981.

Scott, William G. *The Management of Conflict: Appeal Systems in Organizations.* Homewood, Ill.: Irwin-Dorsey, 1965.

Shaiken, Harley. *Wall Street Journal*, 24 September 1984.

"Silicon Valley Labor Organizing Efforts Unsuccessful." *Corporate Times*, February 1985.

Silvestri, Goerge T.; Lukasiewicz, John M.; and Einstein, Marcus. "Occupational Employment Projections through 1995." *Monthly Labor Review* (November 1983): 37–49.

Simpson, Richard L. "Social Control of Occupations and Work." *Annual Review of Sociology* 11(1985): 415–36.

Slichter, Sumner. *Union Policies and Industrial Management.* Washington D.C.: Brookings, 1941.

Slichter, Sumner; Healy, James J.; and Livernash, E. Robert. *The Impact of Collective Bargaining on Management.* Washington, D.C.: Brookings, 1960.

Smith, Bruce L. *The Rand Corporation.* Cambridge: Harvard University Press, 1966.

Smith, Stephen C. "Silicon Valley Days: Human Capital, Profit Sharing, and Incentives in High Tech Firms." Discussion Paper. Washington, D.C.: George Washington University, January 1985.

Solmon, Lewis C., and Beddow, Ruth. "Flows, Costs, and Benefits of Foreign Students in the United States: Do We Have a Problem?" In *Foreign Student Flows*, ed. Elinor G. Barber. New York: Institute of International Education, 1985.

Solmon, Lewis C., and La Porte, Midge A. "The Crisis of Student Quality in Higher Education." *Journal of Higher Education* 57(July-August 1986): 370–92.

"Some GM People Feel Auto Firm, Not EDS Was the One Acquired." *Wall Street Journal*, 19 December 1984, 1, 20.

Stagg, John. "Response to: Changing Technology, Corporate Structure and Geographical Concentration in the Printing Industry." In *Labor and Technology: Union Response to Changing Environments*, ed. Donald Kennedy, Charles Craypo, and Mary Lehman. University Park: Pennsylvania State University, 1982.

"Stakes Are High as H-P Tackles Troubles." *San Jose Mercury News*, 3 March 1985, 1, 16A, 17A.

Steers, Richard M. *Organizational Effectiveness: A Behavioral View.* Santa Monica, Calif.: Goodyear, 1977.

Stinchcombe, Arthur L. "Social Structure and Organization." In *Handbook of Organizations*, ed. James G. March. Chicago: Rand McNally, 1965.

Strauss, Anselm, and Rainwater, Lee. *The Professional Scientist.* Chicago: Aldine, 1962.

Strauss, George. "Professionalism and Occupational Associations." *Industrial Relations* 2(May 1963): 7–31.

Straw, Ronald J., and Foged, Lorel E. "Technology and Employment in Telecommunications." In *Robotics, Future Factories, Future Workers,* ed. Robert J. Miller. Beverly Hills, Calif.: Sage, 1983.

———. "Job Evaluation: One Union's Experience." *ILR Report,* Spring 1982, 24–26.

"Strong Dollar or No, There's Money To Be Made Abroad." *Business Week,* 22 March 1985, 155–62; 246.

Summers, Clyde W. "Individual Protection Against Unjust Dismissal: Time for a Statute." *Virginia Law Review* 62(April 1976): 481–532.

Tisdale, F. "99.44% Security—and Efficiency." *Reader's Digest,* May 1937.

Tolbert, Pamela, and Zucker, Lynne. "Institutional Sources of Change in the Formal Structure of Organizations: The Diffusion of Civil Service Reform, 1880–1935." *Administrative Science Quarterly* 28(1983): 22–39.

Troy, Leo, and Sheflin, Neil. *Union Sourcebook: Membership, Structure, Finance, Directory, First Edition, 1985.* West Orange, N.J.: Industrial Relations Data and Information Services, 1985.

Tsang, Mun Chiu. "The Impact of Overeducation on Productivity: A Case Study of Skill Underutilization of the U.S. Bell Companies." Program Report No. 84–B10. Stanford, Calif.: Institute for Research on Educational Finance and Governance, October 1984.

U.S. Bureau of Labor Statistics. *The Impact of Technology in Five Industries.* Bulletin 2137. Washington, D.C.: U.S. Government Printing Office, 1982.

———. *Earnings and Other Characteristics of Organized Workers.* Bulletin 2105. Washington, D.C.: U.S. Government Printing Office, 1981.

U.S. Congress. House of Representatives. Subcommittee on Science and Technology and Task Force on Education and Employment of the Committee on the Budget, Joint Hearings. *Technology and Unemployment.* 98th Cong., 1st sess., 7, 9, 10, 14, 15, 16, and 23 June 1983.

U.S. Congress. Senate. Joint Economic Committee, Hearings. 98th Cong., 2d sess., 1984. *Climate for Entrepreneurship and Innovation in the U.S.,* pt. 2, pp. 9–107.

U.S. Department of Commerce, International Trade Administration. *United States Trade Performance in 1983 and Outlook.* Washington, D.C.: U.S. Government Printing Office, 1984.

U.S. Department of Education. *Earned Degrees Conferred.* Washington, D.C.: U.S. Government Printing Office, 1976, 1983.

———. *Digest of Educational Statistics.* Washington, D.C.: U.S. Government Printing Office, various years.

———. *Enrollments and Programs in Noncollegiate Schools.* Washington, D.C.: U.S. Government Printing Office, 1976.

———. National Center for Education Statistics. *Digest of Educational Statistics.* Washington, D.C.: U.S. Government Printing Office, 1976–1983.

———. National Center for Education Statistics. *Associate Degrees and Other Formal Awards Below the Baccalaureate.* Washington, D.C.: U.S. Government Printing Office, various years.

U.S. Steel–United Steelworkers. *Agreement Between United States Steel Corporation and the United Steelworkers of America, Production and Maintenance Employees.* 1 August 1980.

Verma, Anil. "Electrical Cable Plant." In *Human Resource Management and Industrial Relations*, ed. Thomas A. Kochan and Thomas A. Barocci. Boston: Little, Brown, 1985.

Vollmer, Howard M., and McGillvray, Patrick J. "Personnel Offices and the Institutionalization of Employee Rights." *Pacific Sociological Review* 3(Spring, 1960): 29–34.

Vroom, Victor. *Work and Motivation*. New York: Wiley, 1964.

Walton, Richard. "From Control to Commitment: Transforming Work Force Management In the United States." Unpublished paper prepared for the Harvard Business School's 75th Anniversary Colloquium on Technology and Productivity, 27–29 March 1984.

———. "New Perspective on the World of Work." *Human Relations* 35, no. 12(1982): 1073–84.

———. "Establishing and Maintaining High Commitment Work Systems." In *The Organizational Life Cycle: Issues in the Creation, Transformation and Decline of Organizations*, ed. John R. Kimberly and Robert A. Miles. San Francisco: Jossey-Bass, 1980.

Watson, Donald Stevenson. *Price Theory and Its Uses*. Boston: Houghton Mifflin, 1972.

Weintraub, E. Roy. *General Equilibrium Analysts*. New York: Cambridge University Press, 1985.

Weisz, William J. "Employee Involvement: How It Works at Motorola." *Personnel*, February 1985, 29–33.

Western Technical Manpower Council of the Western Interstate Commission for Higher Education. *High Technology Manpower in the West: Strategies for Action*. Boulder, Colo: WICHE, January 1983.

"West Germany—Workplace Agreements on New Technology." *European Industrial Relations Review*, no. 90(July 1981): 7–9.

Wilensky, Harold L. "The Professionalization of Everyone." *American Journal of Sociology* 70(October 1964): 137–58.

Wils, Theirry, and Dyer, Lee. "Relating Business Strategy to Human Resource Strategy: Some Preliminary Evidence." Paper presented at the 44th annual meeting of the Academy of Management, Boston, August 1984.

Wilson, Thomas B. *A Study of How It's Done in High-Tech*. Boston: Hay Associates, 1982.

Winpisinger, William. "Job Satisfaction: A Union Response." *American Federationist* 80, no. 2(1973): 8–10.

Woodward, J. *Industrial Organizations: Theory and Practice*. London: Oxford University Press, 1965.

Wright, R.V.L. *Strategy Centers in Contemporary Managing Systems*. Cambridge, Mass.: Arthur D. Little, 1978.

Young, Howard. "The 1984 Auto Negotiations: A UAW Perspective." In *Industrial Relations Research Association Proceedings*, April 1985, 454–57.

Zumeta, William. *Extending the Educational Ladder: The Changing Quality and Value of Postdoctoral Study*. Lexington, Mass.: Lexington Books, 1985.

Index

9–10; scientific and technical, 23, 61,
205–206; specialization in, 13–14;
two-tier, 17, 20, 158–159, 179. *See also*
Labor force; Workers
Work organization and design, in human
resource management model, 187
Workers: blue-collar, 17, 29–30, 31, 33, 93;
displaced, 166; forgotten majority, 17;
monitoring of, 167
Wright, R.V.L., 112, 113

Young, Howard, 169, 181
Young Investigator Awards Program, 71–72

Zucker, Lynne, 18, 189

Contributors

The editors:

Archie Kleingartner is a professor in the Graduate School of Management and associate director of the Institute of Industrial Relations at UCLA. From 1975 to 1983 he served as Vice-President—Faculty and Staff Personnel for the University of California System. He is co-author of *Academic Unionism in British Universities* (UCLA Institute of Industrial Relations, 1986).

Carolyn S. Anderson is a doctoral candidate and assistant to the director of the Institute of Industrial Relations at UCLA. She is co-author of *Building California: The Story of the Carpenters' Union* (UCLA Institute of Industrial Relations, 1982).

The following contributed chapters:

Richard S. Belous is a senior research associate and labor economist with The Conference Board in Washington, D.C. He is the author of *The Computer Revolution and the U.S. Labor Force* (Congressional Research Service, 1985).

John B. Chalykoff is a doctoral student in the Sloan School of Management, Massachusetts Institute of Technology.

Fred K. Foulkes is a professor in the School of Management and director of the Human Resources Policy Institute at Boston University. He is the editor of *Strategic Human Resources Management: A Guide for Effective Practice* (Prentice Hall, 1986).

Everett M. Kassalow is a professor in the School of Urban and Public Affairs at the Carnegie Mellon University and a past president of the Industrial Relations Research Association.

Thomas A. Kochan is a professor in the Sloan School of Management, Massachusetts Institute of Technology, and co-author of *The Transformation of American Industrial Relations* (Basic Books, forthcoming).

Midge A. La Porte is a market research and training consultant for Wishard and Associates, Inc., New Orleans.

David Lewin is a professor in the Graduate School of Business and director of the Industrial Relations Research Center at Columbia University. He is co-author of *Public Sector Labor Relations*, revised edition (forthcoming).

Robert C. Miljus is a professor in the School of Business at Ohio State University and a former member of the governor's task force on union-management cooperation in the state of Ohio.

George T. Milkovich is a professor in the Industrial and Labor Relations School at Cornell University and co-author of *Compensation* (Business Publications, 1986, in press) and *Personnel/Human Resource Management: A Diagnostic Approach* (Business Publications, 1985).

Karl S. Pister is Roy W. Carlson professor of engineering and dean of the College of Engineering, University of California at Berkeley.

Rebecca L. Smith is manager of continuing technical education for Hewlett-Packard, Inc., Cupertino, California.

Lewis C. Solmon is a professor and dean of the Graduate School of Education at UCLA. He is co-author of *Underemployed Ph.D.s* (Lexington Books, 1981).

The following contributed comments:

David G. Ackerman is executive vice-president of the California State Chamber of Commerce in Sacramento and former deputy secretary of the Business, Transportation and Housing Agency for the State of California.

Elinor Glenn is vice-chair of the Employment Training Panel of California and president emerita of the Service Employees International Union, Joint Council, Southern California.

Evelyn Hunt is an attorney with the law firm of Orrick, Herrington and Sutcliffe in San Francisco and co-author of *Academic Unionism in British Universities* (UCLA Institute of Industrial Relations, 1986).

Larry J. Kimbell is a professor in the Graduate School of Management and director of the Business Forecasting Project at UCLA.

Lawrence R. Littrell is corporate director of industrial relations at the Northrop Corporation, Los Angeles.

Frank Matulis is director of human resources at the Interstate Electronics Corporation, Anaheim, California.

Daniel J.B. Mitchell is director of the Institute of Industrial Relations and a professor in the Graduate School of Management at UCLA. He is the author of *Unions, Wages, and Inflation* (Brookings, 1980).

James A. Parrott is an economist and assistant to the president of the International Ladies' Garment Workers' Union in New York. He is co-author of *Massachusetts High Tech: The Promise and the Reality* (High Tech Research Group, 1984).

Harry F. Silberman is a professor in the Graduate School of Education at UCLA and a past chairman of the National Commission on Secondary Vocational Education. He is the editor of *Education and Work* (National Society for the Study of Education, University of Chicago Press, 1981).